St. George Jackson Mivart

An Introduction to the Elements of Science

St. George Jackson Mivart

An Introduction to the Elements of Science

ISBN/EAN: 9783337034610

Printed in Europe, USA, Canada, Australia, Japan

Cover: Foto ©berggeist007 / pixelio.de

More available books at **www.hansebooks.com**

AN INTRODUCTION

TO THE

ELEMENTS OF SCIENCE

BY

ST. GEORGE MIVART, F.R.S.

AUTHOR OF

"TYPES OF ANIMAL LIFE," "ESSAYS AND CRITICISMS,"
ETC.

WITH ILLUSTRATIONS

BOSTON
LITTLE, BROWN AND COMPANY
1894

TO THE DEAR MEMORY

OF MY FATHER

JAMES EDWARD MIVART,

WHO BY HIS TOIL FREED ME FROM SORDID CARES
WHILE HE EVER ENCOURAGED ME TO LOVE
AND WORK FOR SCIENCE,

I DEDICATE

THIS LITTLE BOOK

PREFACE

That a work should be written which might introduce students to the elements of all the sciences, has long seemed to the author of this book a thing to be desired.

Having been unable to find associates for this purpose, he has ventured to undertake it himself.

Strongly convinced that the student ought to be introduced to mental as well as to physical science, the writer has been careful not to omit the elements of psychology, logic, and philosophy, subjects which he thinks have been far too generally neglected.

So far as he knows, it is the first work of the kind, on which account he cannot hope to have altogether escaped errors, not only as to matters of fact, but also in the mode of their presentation. He therefore asks the indulgence of the reader who may deem any subject too fully or too scantily treated of.

Oriental Club,
October 28, 1893.

CONTENTS

CHAP.		PAGE
I.	THE STARTING-POINT	1
II.	MATHEMATICS	6
III.	MECHANICS	40
IV.	PHYSICAL FORCES	84
V.	THE NON-LIVING WORLD	137
VI.	THE LIVING WORLD	185
VII.	MAN	251
VIII.	LOGIC	282
IX.	HISTORY	313
X.	SCIENCE	365

LIST OF ILLUSTRATIONS

		PAGE
1.	Origin of Roman Numerals	8
2.	First Proposition of Euclid	35
3.	Equilibrium Through Opposing Thrusts	42
4.	Centre of Gravity	44
5.	Simple Composition of Forces	47
6.	Complex Composition of Forces	48
7.	Parallel Forces	50
8.	Levers	51
9.	The Pulley	52
10.	The Inclined Plane	53
11.	Angles of Incidence and Reflexion	58
12.	Velocity of a Falling Body	61
13.	The Pendulum	63
14.	Liquid Level Surface	69
15.	The Siphon	80
16.	Relation of Force to Distance	97
17.	Reflexion of Energy	99
18.	Reflecting Mirrors	100
19.	Action of Convex Surfaces	101
20.	Action of Concave Surfaces	101
21.	Refraction	105
22.	Aërial Currents	150
23.	Two Aspects of the Earth	156
24.	Geological Strata	167
25.	Triangular Measurements	177
26.	Horse-Tail	191
27.	Tunicate	192
28.	Animal Cell	195

LIST OF ILLUSTRATIONS

	PAGE
29. Amœba	198
30. Threads of Confervæ	199
31. Protococcus	200
32. Volvox	201
33. Venus's Fly-trap	202
34. Bracken-fern	203
35. Prothallus	205
36. Bean Plant	207
37. Diagram of Ovule	209
38. Leaf of Bryophyllum	211
39. Buttercup	212
40. Radiolarian	214
41. Infusorian	215
42. Hydra	216
43. Diagram of Lion's Eye	220
44. Section of Spinal Cord	226
45. The Right Whale	230
46. The Rhinoceros Viper	231
47. Tadpoles	232, 233
48. The Eft, Amblystoma	234
49. Crayfish	235
50. The Snail	236
51. The Cuttlefish	237
52. Ichthyosaurus	245
53. Plant Mimicry	249
54. Logical Genus and Species	288
55. First Logical Figure	298
56. Second Logical Figure	298
57. Third Logical Figure	299
58. Fourth Logical Figure	299
59. First and Second Moods of First Figure	300
60. Third and Fourth Moods of First Figure	300
61. Fallacious Syllogism	301
62. Dominions and Dependencies of Alexander	321
63. Mediterranean Lands at Beginning of Second Punic War	342
64. The Roman Empire under Trajan	349

ELEMENTS OF SCIENCE

CHAPTER I

THE STARTING-POINT

A LITTLE before the middle of the eighteenth century Buffon published the first volume of his Natural History. It was a wonderful book, and one yet more remarkable for the sagacious or ingenious theories it enunciated with respect to human and animal existence, the past history of this planet, and phenomena of inorganic matter, than for the many descriptions and figures of animals which it contained. Its scope was so great that while it dealt with such matters as the origin of the world and solar system, it also furnished tables representing the probabilities of human life and death—tables which have helped forwards that vast system of life-assurance, to which such a multitude of men and women now owe protection from calamity.

Ever since Buffon's time science has advanced, and a knowledge of it been diffused with greater and greater rapidity; a similar and consequent increase in the comforts and amenities of life being the general result.

As such resulting advantages have become more and more apparent, an increasing appreciation of science itself has naturally followed, while even in the earliest ages of human existence men were compelled to acquire some increase of knowledge with respect to the world about them, in order to preserve their existence during the almost incessant contests between succeeding races of mankind.

The great majority of men still pursue knowledge as a means to attain material advantages of one kind or another, but there is a rapid increase in the number of those who seek it for its own sake.

This is not to be wondered at, since the gratifications which science affords its patient and persevering followers are exceptionally great. Unlike sensuous pleasures, they leave no sting behind them and produce no depressing reaction, but are perennial and untiring. Great is the contrast between the feverish pursuit of gain or the heartburnings of social competition and the calm pleasure afforded by the intelligent contemplation of Nature —a pleasure which can persist unimpaired amongst the otherwise deepening shadows of declining years.

Nor are these advantages beyond the reach of any person of merely normal capacity. The difference between science and ordinary knowledge is no difference of kind but merely one of degree. Science is nothing more than plain reason and common sense used in a methodical manner and applied to the examination of various objects around us, with as much exactness as possible. No one, therefore, who enjoys such knowledge as he has, and feels impelled, through love of it, to acquire more, need feel in any way discouraged. He already possesses the scientific spirit, and nothing but a little patience and perseverance are needed for him, sooner or later, to become a true man of science.

The object of this little book is to assist the student through the very first steps which he must take in such a pursuit, and to introduce him to the means and methods of acquiring scientific knowledge.

The ultimate scope of science is to give its pursuer as correct a knowledge as possible of the nature of each object studied, and the causes which have made it what it is. There are, however, a multitude of things as to which the beginner must at first be content simply to know *that* they are; but none the less his aspirations should always be to know also the *how* and *why* they are whatever they may happen to be.

It is, then, necessary for our purpose to begin with what is most elementary and to suppose the student destitute of any scientific knowledge of the multitude of objects which on all sides solicit his attention. He will probably at first be puzzled how best to begin his study of things so various, as a simple illustration may serve to show.

Let us imagine the would-be student of science walking amongst some partially wooded chalk hills on a brilliant evening in early summer. Birds are singing and insects humming amidst abundant wild flowers. Through a gap in the hills he catches a distant glimpse of the sea, with here and there a sail gleaming in the rays of the setting sun. Entering the wood he follows the margin of a stream which has plainly worn a way for itself, till he comes upon a few sculptured fragments of a ruined abbey which recall to him some erroneous notions that existed when the treasures of the Record Office were still unpublished. Meanwhile dark clouds have gathered in the east, and a brilliant flash of lightning suddenly appears amongst them, and he hears distant thunder while yet the evening star shines calmly near the rapidly setting new moon in the still clear western sky.

What inquiry can we find which shall at one and the

same time equally relate to all the phenomena which will thus have affected the student's senses? What can be common to them all without exception?

Its nature must be wide indeed, since it must be common to things so different. It must be common to animal life and the life of plants, to colour, to sound and to motion, to the sea's waves, the action of running streams, the formation of rocks and hills, and the movements of every breeze; to thunder and to lightning, to earth and to sky, to the sun, moon, and stars, to human history, the progress and decay of institutions, the development of art, and even to the very thoughts which deal with things so various.

Such a combination may well at first seem utterly bewildering, and yet a few very simple reflections may serve to solve the puzzle.

To know anything whatever, is to know that it is distinct from something else. Two marbles, alike in colour and size, shape and weight, are known with perfect certainty to be distinct, though we may not be able to tell one from the other. We recognise them as *two* things of the same kind. Together they form a small group composed of two objects. If now these be held in the right hand, while a third marble, exactly like the other two, is held in the left hand, then the contents of the right hand differs from that of the left simply by being "two" instead of "one"—that is, by a difference of *number*.

But "number" is a property possessed by all the things above referred to, since even thoughts, no less than marbles, differ from each other numerically. Enumeration — accurate enumeration — is necessary for all kinds of knowledge. We may feel things to be hotter or colder, but if we would be accurate we must employ a thermometer and note the degrees registered by it—that is, we must count. We see plainly enough that some things are bigger than others, but if we would be correct we must measure them by some standard, and this again implies counting. It is the same with respect to weight and motion, with respect to our own past history and the past history of mankind. In matters of antiquarian knowledge, bygone periods of time have to be carefully computed, and sometimes the duration of nations and of dynasties. The velocity of winds and waves, the rapidity of the lightning's flash, as well as the seemingly slow revolutions and displacements of the heavenly bodies, have all to be also estimated by counting—that is, by number.

Thus the one thing which alike pertains to everything we know, terrestrial or celestial, material or mental, is "number." It is a certain numerical relation, or rather various numerical relations, since, for example, a nation is one when compared with other nations, but multitudinous when considered with respect to the individuals that compose it. This truth doubtless underlay the system of Pythagoras, who, five hundred years and more before our era, taught that number was the principle of all things.

But the study of that which is thus common to all things, is the study of mathematics; and therefore mathematics, or the science of number, is and must be the most fundamental of all sciences, since it pertains to every other, and no other can be pursued without it. An introduction to the elements of science must therefore begin with an introduction to the elements and principles of mathematics.

CHAPTER II

MATHEMATICS

A STUDY of the first elements and simplest possible principles of mathematics, is then what should first occupy the attention of every would-be student of science. This is absolutely indispensable, since without it no other science is possible; because all of them, without exception, suppose a greater or less acquaintance with it.

An objection, however, has sometimes been thus stated. It has been said that mathematics is a most abstract science, and one, therefore, unfitted to occupy the attention of those whose object is to gain a knowledge of all the concrete, material things about them—the things which they can see, feel, and handle.

Now it is true that the science of mathematics is mainly, and in its simplest branches exclusively, devoted to the study of real or possible numerical relations, apart from the things which bear those relations. Nevertheless common sense shows us at once that numerical relations, or "numbers," can no more exist apart from something which has number, than "weight" can exist apart from something heavy, or "dimension" exist without something or other of a definite size. Numbers, apart from real substantial things, only exist as thoughts, or as the written or spoken signs by which we express numerical relations. But since "numerical relations" have no substantial existence apart from things related, it follows that the science of mathematics —which employs them—ultimately concerns real substantial things themselves, to the study of one aspect of which it is above all devoted.

Everything has number. Larger and smaller groups of similar things differ in number, and we can readily express these differences by spoken or graphic signs to a certain extent. But the limitation of our faculties makes it impossible for us to think or speak of a very extensive series of numbers by entirely distinct symbols. Merely spoken signs we may at once put on one side, as they are entirely devoid of the permanence requisite to enable them to serve for scientific purposes. The art of expressing numbers by means of written signs is called *notation*.

Through the eye, many such numerical symbols are very readily recognisable. Such is, for example, the case with the numerical symbols depicted on dice or cards, and it is conceivable that specially gifted individuals might be able to distinguish and recollect several hundreds of such absolutely different and distinct numerical symbols. That however would, after all, be of but little utility for the study of very high numbers, wherein the most gifted imaginations would soon be reduced to adopt the method employed by ordinary persons. The method usually adopted consists in dealing with numbers in groups, and groups of groups, and groups of groups of groups, and so on, according to some regular system, which for one reason or another has come to be the one adopted.

In the four fingers and thumb of each hand, and the five toes of each foot, man possesses an easy and ready means of incipient enumeration, and the words used by various savage tribes to denote numbers, plainly show that this naturally suggested method has been actually employed by them. Thus, the number "five" is sometimes called a "hand," and "six" is spoken of as "take the thumb"—that is, "begin to make use of the other hand." "Twenty," the number of all the digits combined, is sometimes denoted by the term "a man," and "ten" by "half a man."

The same thing is shown by Roman numerals, where I, II, III, and IIII indicate one to four fingers, while "five" is expressed by a sign representing the thumb upstanding by itself, and the four fingers in a group

FIG. 1.

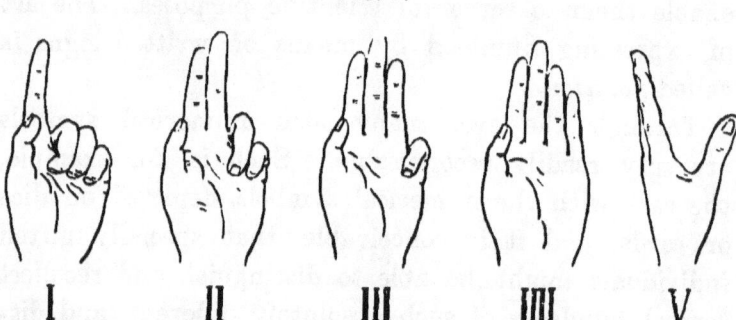

opposite it—V. To express ten, there were the two hands crossed obliquely—X.

Thus an arrangement of numbers in groups of ten naturally suggested itself; and thus ten is the "root" number, or "radix," of the system of counting actually adopted. As written down they form a system of notation, and there being ten symbols (0 1 2 3 4 5 6 7 8 9) to that system, it is called a *decimal* system of notation. But other "root numbers" might have been selected, as we shall shortly see, each giving rise to its own "system of notation."

The Roman numerals, though plainly expressing

numbers, were found comparatively useless for purposes of calculation—purposes which the Arabic symbols have admirably subserved. The absence of numbers being expressed by 0, and the first group of ten (from zero to nine) being expressed by the figures with which we are all familiar, its completion is represented by 10 (or unity and zero combined), and so on with successive groups of ten till the tenth set (90-99) is completed.

Then a third figure is added to the left to denote ten groups of ten (100), while each time such a group is further taken ten times over, it is expressed by the addition of another zero to the right, and it is thus that " ten times, ten times, ten times, ten times, ten times, ten " (or one million) requires to be denoted by the figures 1,000,000. In this way each figure shows its value by the place it occupies. Thus it is that in the symbol 1652392 the figure 2 denotes mere units, 9 the groups of ten, 3 the groups of ten times ten, or hundreds, and so on—or, in other words, the symbol denotes 1 million, 6 hundreds of thousands, 5 tens of thousands, 2 thousands, 3 hundreds, 9 tens, and 2 units.

These truths are, of course, familiar to all readers of this book, though they may not happen to have considered them from the present point of view.

But since our purpose is to introduce the reader to the elements of science, we are bound to act as if exceedingly little were known by him. Thus, to carry out the end we have set before us, we must consider the principles of such elementary processes as addition, subtraction, multiplication, and division.

The definite position of figures, according to their value, greatly facilitates the first of these processes, since by the superposition of figures, thus arranged, we are enabled to add them together as simple units, without taking account of the whole quantities, whereof such figures form part. Thus, in adding together the quantities expressed by 104, 92, and 8, according to this mode, we need take no heed of the three whole numbers as whole numbers, but simply add together their superimposed constituent parts:

$$\begin{array}{r} 8 \\ 92 \\ 104 \\ \hline 204 \end{array}$$

The result of the above simple sum in addition takes the form it does, because as the three superimposed units at the right hand together make 14, we know that the number of simple units is 4, together with one group of ten. When this one group is added to the two superimposed figures which form the second column, the product (because one of them is a zero) is ten groups of ten. But a symbol of that value cannot be written down in the second place, but must appear in the third, which is that set apart for groups of ten times ten. It is therefore carried to the third column, which consists of but a single figure 1, and, being added thereto, makes with it two groups of ten times ten, or 200, so that the result must be 204.

That the results obtained by thus working with mere symbols of abstractions applicable to all things which can be counted, accurately correspond with real relations which exist between substantial things, is, of course, most easily proved. For instance, if we take three parcels of things—*e.g.*, marbles—one of 104, another of 92, and the third of 8, and mix them together, then if we count the whole, thus mixed, we shall find their number to be 204.

The same system serves equally well for subtraction. Thus, if from a group of 204 objects, 20 have to be taken away, we write

$$204$$
$$20$$

Here it is evident (since there is a zero at the right end of the lower number) that no single unit has to be taken away from 204, so the figure 4 must remain unchanged at the right hand of the sum expressing the result. In the second column (which denotes groups of 10) two such have to be taken from zero. This difficulty, as schoolboys know, is evaded by borrowing ten groups of ten from the set of the next higher denomination, then taking two sets of ten from the ten sets thus borrowed, there will remain 8 sets of ten, and 8 will therefore be the second figure of the sum denoting the result. The ten groups of ten, which have been borrowed, have now to be taken away from the third figure, which from its position shows that it denotes groups of ten times ten. This third figure is 2, from which one being deducted, we have, of course, 1 as a remainder, and so we express the process thus:

$$204$$
$$20$$
$$\overline{184}$$

The correspondence of this process with the real relations which exist between substantial things, can again be most simply shown by taking 20 marbles from 204, and counting the number left. In an analogous manner we can (by practical, material tests) establish the correspondence with reality of the other processes of the science of number, which science is the arithmetical part of the great science of mathematics.

Thus by these arithmetical symbols we can elucidate most important results as to real things, without paying attention to anything more than the symbols themselves, till the result sought is attained. If we know that the numbers used refer to marbles or any other set of things, we can freely use them and work out results without thinking of the objects to which they refer till the end of the process. This is of enormous assistance and a prodigious economy of human effort.

There is no special difference between multiplication and addition. Multiplication is the addition of any number to itself a certain number of times over, and a number is said to be multiplied by that number which expresses how many times the former number has to be added to itself. Thus if 10 be multiplied by 2, it has to be added to itself (or taken) twice; if it is multiplied by 9, it has to be taken nine times, and so becomes 90. Nevertheless though multiplication is essentially but a form or mode of addition, practically it is a very different process, and it is one by which we can most clearly see the great convenience of the system of numeration adopted, and of the practice of placing figures in such a way that they express their value by their mere position.

The results of a definite small number of additions of a few small numbers have been calculated for committal to memory according to what we know as *the multiplication table*. This being learnt, two sums, the figures of which have been superimposed in due order of value, can be multiplied together, by the process of multiplying the separate figures which compose such two sums, just as we have seen that two sums can be added together by the addition of the separate figures which compose them.

Let us suppose that the number 2063 has to be multiplied 345 times; that means, either that the number 2063 has to be added to itself 345 times, or that the number 345 has to be added to itself 2063 times, the result of either process being of course the same. This tedious process of addition is avoided by the device of multiplication:

$$\begin{array}{r} \text{Thus } 2063 \\ \text{multiplied by } 345 \\ \hline 10315 \\ 8252 \\ 6189 \\ \hline \text{produces } 711735 \end{array}$$

In this way we quickly arrive at the result of adding 2063 together, first 5 times, then 40 times, and lastly 300 times. If we wish to test the first process, we must see that 2063 added together five times produces the same result as the multiplication of that sum by the number 5, thus:

$$\begin{array}{r} 2063 \\ 2063 \\ 2063 \\ 2063 \\ 2063 \\ \hline 10315 \end{array}$$

In obtaining the same result by multiplying, we see by the multiplication table just referred to, that 5 times 3 are 15. We therefore set 5 down at the extreme right, but reserve the single group of 10 (out of the 15) till we see how many groups of 10 will be produced by the next step of multiplication. Our table tells us again that 5 times 6 are 30, and the 6 there multiplied (since it stands in the second place from the right) denotes 6 groups of 10, so that 30 resulting from its multiplication by 5 means 30 groups of ten, to which we add the one group of ten reserved out of the 15 units—making 31. Of this 31 we again write down 1, reserving the 3 to be added to the sum of next higher value. But the figure which stands next is a cypher or zero, and 5 times nothing is, of course, nothing, so all we have next to set down is the 3 we previously held in reserve. Lastly comes the fourth figure 2 (denoting two groups, each of ten times ten times ten), and this, when multiplied by 5, becomes 10, and so we arrive at the result, "ten thousand three hundred and fifteen," which sum we also obtained by simple addition. Then comes the addition of 2063 to itself 40 times, which is effected by multiplying the 2063 by a 4 which stands in the second place, and therefore denotes not four units but 4 groups of ten. We thus learn by a brief process that 2063 added to itself forty times over comes to 82,520. The zero is not indeed written down in the process because we have now nothing to do with mere units, the 4 used in multiplying denoting, as before said, only groups of ten.

Similarly we can quickly see the result of adding 2063 to itself 300 times by multiplying the former sum by a figure 3 standing in the third place, which is that set apart for denoting what the number of hundreds may be. We thus see that the addition of this sum to itself 300 times, amounts to the number 618,900. The two cyphers are not written down because the multiplying figure being one of hundreds, we have no longer anything to do with sets of ten only, and still less with mere units. Having now these three products of the addition of 2063 to itself 5 times, 40 times, 300 times, we have but to place them one under the other and add them up thus :

$$\begin{array}{r} 10315 \\ 82520 \\ 618900 \\ \hline 711735 \end{array}$$

And thus we know that the result of the addition is equivalent to the adding 2063 to itself 345 times.

When a number is multiplied (added to itself) its own number of times, as, *e.g.*, 5, five times, or 9, nine times, or 1000, a thousand times, the product is called the square of each such number. The number which, by so multiplying itself, makes that product, is called the *square root* of that same product whatever it may be.

When the square of a number is again multiplied by that number, the product is called a *cube*, and the original number is the *cube root* of such product.

The square, or the cube, of a number may be represented by a small figure, which is called an *index*, placed on one side above the number squared or cubed. Thus the indices 2 and 3 thus placed with respect to 4, will be 4^2 and 4^3; and these two symbols respectively indicate 4 squared and 4 cubed—which are, of course, 16 and 64. The symbols $\sqrt{}$ and $\sqrt[3]{}$ indicate respectively the square and cube roots of any numbers, and thus $\sqrt{16}$ is 4 and the $\sqrt[3]{64}$ is also 4.

The number of times any quantity is thus multiplied by itself is called its "power." Thus 2^7, or 2 raised to the seventh power, is 128, and, of course, any number may be raised to any power.

The process of *Division* is a form of subtraction, as the process of multiplication is a form of addition. It is a process which shows us, by the aid of the multiplication table, how many times one number may be contained in another. Thus, *e.g.*, we see that 2 is contained 8 times in 16, because twice 8 are 16. We express it familiarly thus:

$$2 \overline{)16}$$
$$8$$

But the same result is arrived at (and its correctness, if need be, proved) simply by a repetition of the process of subtraction, thus:

$$\begin{array}{r} 16 \\ 2 \\ \hline 14 \\ 2 \\ \hline 12 \\ 2 \\ \hline 10 \\ 2 \\ \hline 8 \\ 2 \\ \hline 6 \\ 2 \\ \hline 4 \\ 2 \\ \hline 2 \end{array}$$

which demonstrates that it consists of 2 eight times taken.

In dividing large numbers by one another, we make use of a device analogous to that of multiplication, beginning, however, with the other end of the series. We begin in this way, because, in division, we have first to do with symbols expressing the highest value concerned, the simple units coming last.

Thus if, *e.g.*, 40,925 be divided by 362, we then see both how many times the lesser number is contained in the greater and what still lesser number remains as a residue. Making use of the multiplication table and writing down the process in the usual way, we have :

$$
\begin{array}{r}
362)\overline{40925}(113 \\
362 \\ \hline
472 \\
362 \\ \hline
1105 \\
1086 \\ \hline
19
\end{array}
$$

which shows us that the lesser number is contained 113 times in the greater number and that 19 units remain over.

As most readers of this book of course know, there are symbols, not only for numbers representing units, but also for parts of units or *fractions*: such as $\frac{1}{2}$ (a half), $\frac{1}{5}$ (a fifth), $\frac{7}{9}$ (seven-ninths), &c.

The figure below the line is called the *denominator*, because it indicates what proportion (or "denomination") of a whole number it is; while the figure above the line is called the *numerator*, because it indicates of how many units of that "denomination" the fraction in question consists.

Here the root number 10 comes again into play in a way analogous to that before mentioned.

A tenth part being written $\frac{1}{10}$, the tenth part of a tenth part, is expressed by adding a zero to the right, $\frac{1}{100}$, and so on indefinitely.

This fact has suggested a further development of the system previously described.

We saw that in any series of figures expressing a number, the figure at its right extremity signifies units, while each succeeding figure to the left expresses a higher power of ten. Now evidently we may also add figures to the right of the figure expressing units, and then each succeeding figure will express a decreasing power of ten, just as well as a fraction will, and we place a point to indicate the spot where this decrease begins. Thus one and one-tenth, which we may write as $1\frac{1}{10}$, may be equally expressed by 1.1, and similarly:

$1\frac{1}{100}$ by 1.01
$1\frac{1}{1000}$ by 1.001, and so on.

Thus 892.35 means, 8 groups of ten times ten units, 9 groups of ten units, 2 units, 3 groups of tenths of units, and 5 groups of hundredths of units.

So far we have dealt only with enumeration according to the radix 10—the decimal system of notation. But, as before said, the employment of this radix simply arose from the number of our fingers and toes. We may take any number as a radix, but if we had had six digits on each hand, the radix we should have taken would no doubt have been 12, which would have constituted a duodecimal system of notation. This would have been a more convenient one, since 12 can be divided by 2, 3, 4, and 6, while ten can only be divided by 2 and 5.

To express numbers duodecimally, *i.e.*, when twelve is taken as the radix, we require two more symbols to express 10 and 11 respectively by single figures. If we represent 10 by the symbol ⁊, and 11 by ℮, then, of course, groups of twelve will need to be represented by two figures, and groups of twelve times twelve by three figures—groups of twelve always taking the place of the groups of ten in the ten radix system. They may be expressed as follows.

Numbers expressed in the radix of 10.	Numbers expressed in the radix of 12.
1	1
⋮	⋮
9	9
10	⁊
11	℮
12	1
13	11
⋮	⋮
18	16
19	17
20	18
21	19
22	1⁊
23	1℮
24	20
25	21
⋮	⋮
60	50
⋮	⋮

Numbers expressed in the radix of 10.	Numbers expressed in the radix of 12.
100	84
⋮	⋮
144	120 (12 times 12)
⋮	⋮
1728	1440

But instead of 10 or 12, we might make use of a binary system of notation, that is, we might take 2 as the radix. Then to express numbers according to such a system, an additional figure would have to be added to the right for every increase of 2, as follows:

Radix of 10.	Radix of 2.
0	0
1	1
2	10
3	11
4	100
5	101
6	110
7	111
8	1000
9	1001
10	1010
11	1011

Thus according to this system every time the symbol one is moved one space to the left, its value is doubled—as we see above (in the radix of 2), where 10 is twice 1; 100 twice the value of 10, and 1000 twice the value of 100.

If we square (multiply by itself) this last number

11, in the radix of two, we produce the following result:

$$
\begin{array}{r}
1011 \\
1011 \\
\hline
1011 \\
1011 \\
0000 \\
1011 \\
\hline
1111001
\end{array}
$$

This may be proved to be right by analysing the product and comparing it with a similar analysis of 11 times 11 in the system of the radix 10.

Radix of 2.				Radix of 10.
1	=	1	=	1
1000	=	2^3	=	8
10000	=	2^4	=	16
100000	=	2^5	=	32
1000000	=	2^6	=	64
1111001				121

Now 1111001 is 11 multiplied by itself according to the radix of 2. But 121 is also 11 multiplied by itself according to the radix 10:

$$
\begin{array}{r}
11 \\
11 \\
\hline
11 \\
11 \\
\hline
121
\end{array}
$$

Various symbols are used to denote, not quantities, but certain relations between them. Thus the symbol = shows and indicates "equal to." Other symbols are useful as follows:

+ (plus, or added to)   − (minus, or taken from)
× (multiplied by)   ÷ (divided by)

Thus $6 + 3 = 9$. $9 - 4 = 5$. $3 \times 2 = 6$. $10 \div 5 = 2$.

These signs are of special use in algebra, as we shall shortly see.

This explanation of the fundamental conceptions and simplest practices of the arithmetical part of mathematics will suffice for our purpose, which is but to show (1) what are the truths at once the simplest and the most universal, because applicable to all objects which can be enumerated, and (2) to make it clear that by working with such symbols we can arrive at definite results which correspond (with perfect exactness and certainty) to real relations existing between objects of all kinds.

For information respecting the manifold, complex, and most ingenious processes and devices whereby the labour of counting and calculating is lightened, the reader is referred to explicit treatises on the rules and practice of arithmetic.

Arithmetic concerns itself with definite numbers, whole or fractional, and each symbol it employs denotes some quantity or other. But, since the Middle Ages, a much wider and more searching branch of mathematics has been widely cultivated—namely, *Algebra*.

Each arithmetical operation applies only to certain numbers, but each algebraic operation is, at one and the same time, good for all numbers, whole or fractional—

i.e., for indefinite quantities of all kinds, known and unknown.

Algebra is a further extension of that process of abstraction which is employed in arithmetic. In arithmetic we use symbols to denote definite quantities of undefined things. Thus, we use 7, 9, and 12 to denote such *definite quantities* of any kind of things whatever. In algebra we use symbols to denote *undefined quantities* of undefined things. An algebraic statement—*e.g.*, $a + a = 2a$—applies to any possible quantities or any possible or impossible things. That economy of human effort which is effected by arithmetic is, as before said, carried to enormously greater extent by algebra.

Such indefinite quantities as are treated of in algebra are represented by letters. It is usual in elementary algebra to represent definite and constant quantities by the first letters of the alphabet, a, b, c, d, &c., and to represent quantities which are variable, are under investigation, and have to be determined, by the last letters of the alphabet, z, y, x, w, &c.

Capital letters, Greek letters, and various other symbols, are used to denote quantities according to circumstances. As to symbols denoting *relations between quantities*, in addition to those lately referred to as of special use in algebra, the following may be added out of a variety of other ones: The sign $>$ between two quantities signifies that the quantity expressed on the left hand of the sign is greater than that on its right, as $a > b$ means that a is greater than b.

Similarly, $a < b$ means that the right-hand quantity (here a) is less than that on the left hand.

When letters representing quantities are enclosed in a bracket, or have a line drawn over them, each of these symbols signifies that such quantities are to be taken as one whole, or collectively. Thus, instead of writing $a \times x + b \times x^* + c \times x$, we may write $(a+b+c)x$, the x being placed outside the bracket. This means that the whole of the quantity contained within the bracket is to be multiplied by x. We may express the same thing thus :

$$\overline{a+b+c} \times x.$$

The sign ∴ means *therefore*, and the sign ∵ means *because*.

Numbers as well as letters may be used in algebra. Thus, $2a + 4b - 3c$ would represent 18, if $a = 2$, $b = 8$, and $c = 6$, for twice 2 are 4, and four times 8 are 32, and those two numbers added make 36, while if three times 6, which is 18, be taken away (as the minus sign indicates) there remains 18.

Such numbers or letters prefixed to symbols of quantities, are called " coefficients."

Fractions are written as in arithmetic : *e.g.*, the number a of any denomination, b is written $\frac{a}{b}$.

If the denominator be a power of any quantity—as $\frac{1}{a^1}$, or $\frac{1}{a^3}$, or $\frac{1}{a^n}$, then such a quantity may be expressed in algebra by what is called a " negative power " of such quantity, that is by a corresponding index* with the negative sign before it.

Thus, $\frac{1}{a^1}$ may be † written a^{-1}.

,, $\frac{1}{a^3}$,, ,, a^{-3}.

,, $\frac{1}{a^n}$,, ,, a^{-n}.

(n, of course, standing for any number.)

---

* See *ante*, p. 15. † See bottom of p. 25.

Points are made use of to denote proportion, thus— $A:b::c:d$ signifies that a bears the same proportion to b as c does to d.

There are certain evident truths or simple axioms which the student must bear in mind, thus:

1. If equal quantities be added to equal quantities, the sums will be equal.

2. If equal quantities be taken from equal quantities, the remainders will be equal.

3. If equal quantities be multiplied by the same, or equal quantities, the products will be equal.

4. If equal numbers be divided by the same, or equal quantities, the quotients will be equal.

5. If the same quantity be added to and subtracted from another, the value of the latter will not be altered.

6. If the same quantity be used both to multiply and divide any quantity, the value of the latter will not be altered.

It may be useful here to observe that the signs + and −, have in our day acquired an exceedingly wide signification. They are now used to denote all sorts and kinds of opposition, not only with regard to quantity, but opposite relations of all kinds—time, space, velocity, or any other property. So if + be applied to any operation, direction, or quality then − will denote the inverse operation, the opposite direction and the most opposed quality. Thus if + refers to North, to increase, to stability, &c., − will denote South, decrease, instability, &c.

A specially algebraic illustration may be derived from the fact that since a means a times any unit, a^1 which is a taken one time, signifies 1 multiplied by a, and since the negative sign denotes the opposite operation a^{-1} must denote 1 divided by a.

Owing to the extremely wide significance these symbols have acquired, it is now also very common to use for either + and − the term "*sense*" instead of "*signs*," *i.e.*, "positive sense" for +, and "negative sense" for −.

Algebraic addition is partially like, yet in some respects different from, arithmetical addition.

If similar quantities have different signs, then their difference must be taken into account when they are added together.

$$\begin{array}{r} \text{Thus, if to } a + b \\ a - b \text{ be added,} \\ \hline \text{the result is } 2a \end{array}$$

because the $+ b$ and the $- b$ neutralise each other, while each a remains.

Similarly, if there are co-efficients, their differences must be taken into account. Thus, by adding together

$$\begin{array}{r} 3a^2 + 4bc - e^2 + 10 \\ -5a^2 + 6bc + 2e^2 - 15 \\ -4a^2 + 6bc - 10e^2 + 21 \\ \hline -6a^2 + 16bc - 9e^2 + 16 \end{array}$$

we obtain the above as the result.

Though convenient, it is not necessary, to write the same letters over each other, but similar quantities must be collected together in stating the result, thus:

$$\begin{array}{r} ax^2 + cz + by^2 \\ 2bx^2 + cy^2 + az \\ - z + x^2 + y^2 \\ \hline ax^2 + 2bx^2 + x^2 + by^2 + cy^2 + y^2 + cz + az - z \end{array}$$

This may be written more shortly and conveniently by

using brackets, since x^2 is taken a times, $2b$ times and once; and y^2 is taken b times, and c times and once; and z is taken c times, and a times and minus once; we may evidently write it with the result as follows:

$$(a + 2b + 1)\,x^2 + (b + c + 1)\,y^2 + (c + a - 1)\,z.$$

Thus the addition of algebraical quantities is performed by connecting those that are unlike with their proper signs, and collecting those that are similar into one sum.

In algebraical subtraction, on the other hand, we have to change the sign, and then proceed as in addition.

The reason of this change of sign is best seen by an example; and the reader must bear in mind the fifth axiom before given.

Let us suppose that from any quantity a, there has to be subtracted the quantity $b - c$. Now if we subtract b from it (which would be expressed thus, $a - b$), we shall have subtracted too much, because the quantity to be subtracted was not b, but only whatever might be left of b after c had been taken away from it. It was not the whole sum of b, but only b diminished by c, or $b - c$, which had to be taken from a: therefore evidently the operation will be completed by adding c to the too much diminished sum, $a - b$.

Thus we have $a - b + c$, and so we have come to change the sign before c from $-$ into $+$. It follows that, to subtract $b - c$ from a, we must change the signs and add.

Therefore in order to subtract from

$$+ a$$
$$\text{the sum } + b - c$$

we must change the signs of the quantities to be subtracted; thus:

when we have for result $\overline{+a-b+c}$, which is correct.
$$\begin{array}{r}+a\\-b+c\text{ and then add};\end{array}$$

Any quantity preceded by the sign $-$ is a negative quantity.

On the rational principle of our language, that "two negatives make an affirmative," to take away a negative quantity from any other quantity is really to add to that second quantity. Thus if 5 has been taken from 12, so that 7 remains, and then that operation be *negatived*, that amounts to adding 5 again to the 7 and so restoring the original number 12.

Similarly, if both 2 and 3 have to be taken from 10, we may write it $10-(2+3)=5$. But if no bracket be used, we must of course change the sign in order to show that both 2 and 3 are taken from 10, and write it, $10-2-3=5$; $10-2$ being 8, and $8-3$ being 5.

Suppose we have to subtract $+by$ from $+3ax$, the difference is obviously $3ax-by$; and thus the sign before by is changed; but if instead of the positive quantity $+by$ we have to take the negative quantity $-by$ from $+3ax$, the result then must be $3ax+by$.

This may seem at first paradoxical to some readers, but to take away a negative (*i.e.*, to subtract a diminution) is evidently, in fact, to make an addition. To cause a man to cease to have no hat is, of course, to cause him to have one.

The above statement may be made more plain by the fifth axiom, for if we both add and subtract the same quantity to and from $3ax$, then, of course, $3ax$ will remain unchanged and as it was. Now if we accordingly add to and take from it cy, we shall have $3ax+cy-cy$, which is simply the same quantity as $3ax$. Let us then take $-cy$ from both, and the result must (according to the 2nd axiom) be the same in each case. But if we take $-cy$ from $3ax + cy - cy$, the result, of course, is $3ax + cy$. Therefore, if we take $-cy$ from $3ax$, the result must also be $3ax + cy$.

$$\begin{array}{l} \text{From } ax^3 - bx^2 + x \\ \text{Take } px^3 - qx^2 + 2x \\ \hline \text{The result or difference} = (a-p)x^3 - (b-q)x^2 + (1-2)x. \end{array}$$

Here we have the quantity x^3 twice repeated, each time with a different coefficient, and the coefficient $+p$, has to be subtracted from $+a$, the result necessarily being $ax^3 - px^3$, which may be written $(a-p)x^3$.

Of the two squares of x, the negative coefficient $-q$, has to be taken from $-b$; we must then, as before, change the sign of q for subtraction, and so we have $-bx^2 + qx^2$, and this may be expressed in a bracket $-(b-q\,*)\,x^2$. Finally the simple quantity x and the coefficient 2 have to be subtracted from the quantity $1x$ (since x standing alone is one x) and so we have $(1-2)x$.

In algebraic multiplication the explicit sign of that process (×) is often omitted, and any two letters written with only a point between them $(a.b)$, or merely side by side (or ab) mean (as in arithmetical multiplication) that a has to be taken b times, or that b has to be taken a times.

If the quantity which is to be multiplied (or the multiplicand), and the quantity by which it has to be multiplied (or the multiplier) have both the same sign (both + or both −), then the result must have the

---

* The portion b remaining after q has been subtracted from it, or $(b-q)$, being of course equal to that produced by the subtraction of the whole of b, followed by the addition of q, or $-b + q$.

positive sign (+) prefixed to it. If they are unlike, then the result must have the negative sign (−) before it. Thus $+ a \times + b = + ab$; $- a \times + b = - ab$; $+ a \times - b = - ab$, and $- a \times - b = + ab$.

The reason of this is very simple, $+ a \times + b = + ab$, because a has to be taken positively b times; and $- a \times + b = - ab$, because the sum $- a$ has to be taken b times, as is expressed by the result. But $+ a \times - b$ also $= - ab$, because multiplication being, as before said, essentially the same as addition, multiplying a by b is the same thing as adding a to itself b times. Now, in this case, a has to be added to itself $- b$ times, which is of course less than once, or, in other words is really subtraction. Thus, in this negative case, a has to be subtracted from itself b times, and (as we have seen in subtraction) the sign must be changed and so a subtracted from itself b times is, and must be $- ab$. Lastly $- a \times - b = + ab$, because here, on the same principles as in the last case, $- a$ has to be subtracted from itself b times. Therefore it is $- ab$ which has to be subtracted; but, as we have seen, to subtract a negative quantity is the same thing as adding a positive one, and therefore subtracting $- ab$ is the same thing as adding $+ ab$, and therefore $- a \times - b = + a\, b$.

It may be useful to note the three following examples of multiplication:

$$\begin{array}{r} a + b \\ \text{multiplied by } a + b \\ \hline a^2 + ab \\ + ab + b^2 \\ \hline a^2 + 2ab + b^2 \end{array}$$

Therefore here we see what is the square of the quantity $a + b$ or $(a + b)^2$.

Therefore also, the square root of the product, or,

$$\sqrt{a^2 + 2ab + b^2} = a + b.$$

Again, if $a + b$ be multiplied by $a - b$

$$\begin{array}{r} a^2 + ab \\ -ab - b^2 \end{array}$$

The product equals $a^2 \qquad - b^2$ since $+ab$ and $-ab$ neutralise each other.

If we multiply $1 - x + x^2 - x^3$
by $1 + x$

$$\begin{array}{r} 1 - x + x^2 - x^3 \\ + x - x^2 + x^3 - x^4 \end{array}$$

The result is $1 \qquad\qquad - x^4$

The other quantities neutralise each other.

In algebraical, as in arithmetical, division, the question is to determine how many times one quantity, "the divisor," may be contained in another, "the dividend," which is equivalent to finding out what quantity multiplied by the divisor will produce the dividend.

Thus to divide ab by a, is to determine how often a must be taken to make up ab; that is, what quantity multiplied by a will give ab, and this we know to be b.

The signs change as they do in multiplication, and for the same reason. If the divisor and dividend have like signs the quotient is $+$; but $-$, if they have unlike signs.

Thus $-ab$ divided by $-a = b$; because $-a$ (the divisor) and $+b$ (the quotient) if multiplied together give $-ab$ (the dividend).

If we divide $a^2 + 2ab + b^2$ by $a + b$ (its square root as we just saw*) the quotient must equal the divisor, thus :

$$a + b \overline{)\, a^2 + 2ab + b^2} \, (a + b$$
$$ a^2 + ab $$
$$ ab + b^2$$
$$ ab + b^2$$

Sometimes quantities may be continued on indefinitely, as when 1 is divided by $1 - x$.

$$1 - x \overline{)\, 1} \qquad (1 + x + x^2 + x^3 \text{ &c. &c.}$$
$$ 1 - x$$
$$ + x$$
$$ + x - x^2$$
$$ + x^2$$
$$ + x^2 - x^3$$
$$ + x^3$$
$$ + x^3 - x^4$$
$$ + x^4, \text{ &c. &c.}$$

The foregoing observations must suffice as a first introduction to the principles of algebra, as a branch of science replete with the most beautiful, complex, ingenious, and far-reaching processes, whereby alone many calculations are made possible, or the labours of investigation lessened, while the results arrived at have extraordinary accuracy. Though for these purposes we may employ not only purely imaginary, but even

---

* See *ante*, p. 31.

impossible quantities, yet the results of the facts and laws thereby discovered (like those of arithmetic) correspond with the facts and laws of real or possible existences. They express abstract truths which have real applications or would have them could the impossible conditions sometimes supposed really exist. Thus even the absurd and impossible quantity expressed by the symbol $\sqrt{-x}$ has its relation with reality. It is really impossible in itself, since there is no quantity which, being multiplied by itself, gives a negative product. Yet it has its relation with reality, inasmuch as it can be used as if it were a real quantity, and all the laws and relations relating to real quantities can be applied to it. Thus:

$$\sqrt{-x} \times \sqrt{-x} \times x = -x^2.$$

Thus we may investigate laws concerning space, motion, pressure, &c., apart from certain conditions which in fact always exist, but which may be temporarily disregarded.

The results so arrived at will be absolutely true, though of course they will not correspond with the phenomena of the world about us, till we take into consideration the conditions which before had been purposely left out of the calculation. These being correctly restored and added, the results will correspond with the realities of experience.

The truths and processes of algebra may be tested by selecting any numbers as representatives of the algebraic symbols (which latter are valid for all numbers) and treating them similarly. This translates the results into arithmetic, and arithmetical results may then be tested by experiments with corresponding numbers of material bodies.

Thus as an example of the correspondence of algebraic truths with arithmetical ones, let us, for example, represent a by 2 and b by 3.

We now know that $a + b$ multiplied by $a + b$ equals $a^2 + 2ab + b^2$.

Similarly $(2+3) \times (2+3) = 2^2 + 2(3 \times 2) + 3^2$.

For $2 + 3 = 5$ and $5 \times 5 = 25$.

Also $2^2 = 4$, while $2(3 \times 2) = 12$ and $3^2 = 9$.

And $4 + 12 + 9 = 25$.

The sciences of numbers and quantity apply, as before said, to all things without exception. A less universal branch of mathematics relates to all things with length, breadth, and thickness. This is geometry. A brief account of its simplest truths will serve to conclude our introduction to the elements of mathematical sciences.

The simplest way of introducing the reader to the elements of geometry will be to explain a proposition of Euclid. The first of his propositions solves the problem how to draw an equilateral triangle (*i.e.*, one all the three sides of which are equal) upon a given straight line of a certain definite length.

To do this we must take the following premisses for granted:

1. That a straight line may be drawn from one point to another;

2. That a circle may be drawn from any given centre at any practicable distance from it;

3. That a circle is such a figure that all straight lines drawn from its centre to its circumference (*i.e.*, to the single line which bounds it) are equal to one another; and,

4. That things which are equal to the same thing are equal to each other.

These truths (which are some of the *definitions* and

axioms* of Euclid) being granted, the problem is solved as follows, and the reader will see that the solution is absolutely certain for any such possible triangle.

Let us first draw a line (as from premiss 1 we can) from a point marked A to a point marked B, and let this be the given line whereon the equilateral triangle is to be drawn.

Now taking the point A as a centre, let us describe round about it (as from premiss 2 we can) a circle, the

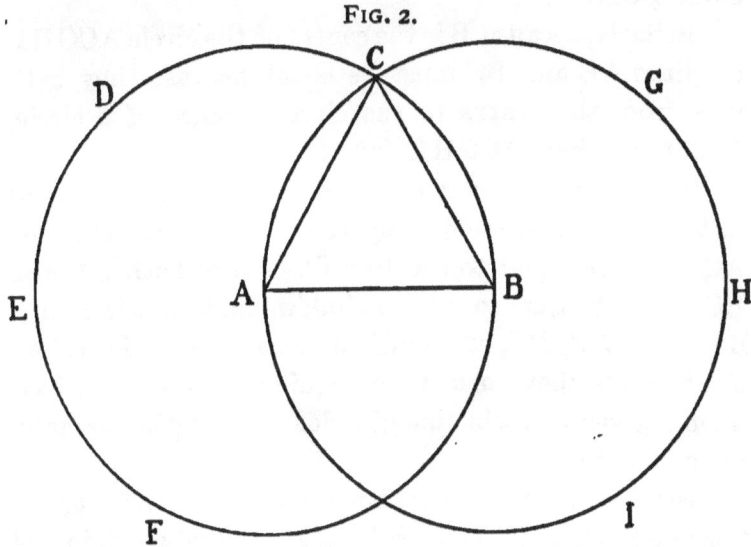

Fig. 2.

circumference of which shall pass through the point B forming the circle BCDEF. Next, taking the point B as a centre, let us describe round about it a circle, the circumference of which shall pass through the point A, forming the circle ACGHI. From the point C, where these two circles intersect, let us draw two straight lines, one from C to A and the other from C to B, as we may do from premiss 1.

---

* See *ante*, p. 25.

Then the triangle ACB will be the triangle required, *i.e.*, it will be an equilateral triangle drawn upon a given straight line of a certain definite length—namely, from A to B.

This is and must be so, for the following reasons:

Since A is the centre of the circle BCDEF, it follows, from premiss 3, that the two lines AB and AC must be equal, since they are both lines which pass from the centre to the circumference of the same circle, *i.e.*, the circle BCDEF.

Similarly, because B is the centre of the circle ACGHI, the lines AB and BC must be equal, because they both pass from the centre to the circumference of a circle, *i.e.*, of the circle ACGHI.

But we have already seen that the line AC is equal to the line AB, therefore (by the 4th premiss) the line AC must be equal to the line BC—since both AC and BC are each equal to AB. It follows then that the three lines AB, AC, BC, are equal to one another. Therefore the triangle they form is an equilateral one described upon a given straight line of a definite length—namely, upon the line AB.

Proofs analogous to the above, support all the propositions of Euclid, and the results are absolutely certain and true. In nature, the properties of bodies as regards their occupation of space—or, as it is called, their "extension"—correspond as accurately with the laws of geometry as their material conditions render possible. Obviously the lines and surfaces which can be made in some substances are less definite and exact than those which can be formed in others, and in no substance can lines and surfaces of ideally perfect straightness, &c., be produced.

But such deviations from ideal perfection, in no way

invalidate the absolute truth of the determinations of geometry themselves, which are more accurately conformed to, the more the nature of any material renders it able to approach more nearly to the perfection desired.

Geometry arose through desires and efforts to measure land accurately, and the properties of angles and triangles actually serve this process now. One of the most useful properties of triangles consists in the fact that two of them, however different in size, are in other respects exactly similar to each other if the angles of one are severally equal to the angles of the other. It is by the aid of such considerations that many of the most important and prodigious scientific measurements have been effected.*

Euclid's work treats not only of lines, angles, triangles, circles, &c., but of the geometrical properties of solid figures of several different shapes.

Greek geometers occupied themselves, in a purely speculative manner, with the different methods in which a circular cone may be cut. The investigation of the various kinds of curves which may be produced at the edge of such a cone by cutting across† it in different directions, constituted the study known as "Conic Sections." The importance of these investigations will become clear when we have to consider falling and other movements of various bodies.

Very many geometrical propositions which were long thought incapable of investigation and solution save by the method proper to geometry, were subsequently found capable of more convenient treatment by the aid of algebra, a change which has produced most important results in the study of astronomy.

---

* See *post*, p. 177.   † See *post*, p. 65.

Even the beginner may see how, in some instances, a geometrical proposition may be more conveniently treated algebraically. Thus, *e.g.*, there is one * which declares that if a right line be divided into any two parts, the square of the whole line is equal to the square of the two parts together with twice the product of those parts.

Now evidently this is equivalent to saying that if we take a to represent one of the two parts into which the right line is divided and b to represent its other part, then the square of the whole line is equal to the squares of a and b together with twice the product of a and b, and this must be $a^2 + 2ab + b^2$, which, as we saw before,† is the result of multiplying $a+b$ by itself, or in other words is equivalent to $(a+b)^2$.

Of late years a converse process has taken place, and various algebraic processes have been converted into geometrical demonstrations, which, as less highly abstract, are more readily apprehensible.

By a number of elaborate processes (which, however elaborate, are essentially similar to and wholly based upon the elementary matters herein pointed out) the most varied properties of objects may be investigated, including complex reciprocal relations of increase, decrease, and variation. When two quantities vary, they may do so equally or in different proportions or ratios. When one quantity varies with another, it is said to be a *function* of the latter. There are many other divisions of the science, whereof one is known as the *Differential Calculus* and deals with computations concerning the rates of change between quantities, while another, called the *Integral Calculus*, passes from the relation between such rates back to the relations existing between the changing

---

* Euclid, Book II., Proposition IV.   † See *ante*, p. 30.

quantities themselves. With such matters the highest branches of mathematics are concerned, but they are, of course, quite beyond the range of an introduction to the elements of science.

For information concerning all but the rudiments of mathematics, the reader is referred to the various works specially devoted to the teaching of that science.

But before concluding this chapter we desire again to insist on the correspondence which exists between the truths of mathematics, of whatever order, and the properties of real substantial things. The truths of geometry are made evident to the eye by diagrams and to the mind by reasoning. The truths of algebra may, as before said,* be tested by taking certain numbers as exponents of the algebraic signs, and so reducing algebra to arithmetic, while the truths of arithmetic may be demonstrated by the seeing and handling of corresponding numbers of real material bodies. We may now pass on to the study of elementary truths, second only to mathematics in the universality of their application.

---

* See *ante*, p. 33.

CHAPTER III

MECHANICS

Having, in the preceding chapter, considered the first elements of that branch of science which is the most universal of all (since it relates to all things which can be counted), we may now proceed to make some acquaintance with the science which treats of the next most general and obvious properties and powers of the substances and bodies which make up the world about us, namely, with the science of Mechanics.

We have hitherto only been concerned with ideas of *number*, *space*, and *direction*, but in mechanics we shall be compelled to deal with *time*, *motion*, and *force*.

All bodies known to us may be roughly arranged in three groups: (1) a group of *solids*, such as pieces of stone, metal, wood, &c.; (2) a group of *liquids*; and (3) a group of substances more or less like the air we breathe, or like gas, and which are thence termed *aëriform* or *gaseous*. Liquids and gases are also classed together as *fluids*, on account of the ready mobility of both.

Any of these three kinds of bodies may be apparently in a state of rest, or obviously in motion, and the study of the various circumstances which attend these two conditions of such bodies, constitutes the "Science of Mechanics."

As every one knows, solid bodies and liquid substances, when unsupported, fall to the ground, such apparent exceptions as balloons, &c., not being really exceptions— as will be seen later on. We say that bodies are "heavy," and that it is their weight which makes them fall. But this "weight" of theirs also causes them to press with greater or less force, so to speak,* on whatever supports them. Many things (intentionally or unintentionally) are thus so pressed and squeezed that they become flattened out—thus making such pressure evident to our senses.

In this way it becomes plain that most (if not all) things tend to fall, not only to the general surface of the ground but as much deeper as circumstances may render possible—as water, stones, &c., will be sure to fall to the bottom of the deepest excavation, unless arrested by something which checks such fall and sustains the falling bodies. We may then fairly assume that whatever tends to fall, tends to fall towards the centre of the earth.

Therefore everything on its surface which appears to be (and for us practically is) in a state of rest, is really tending to move and is only prevented from actually moving by some other object which checks its progress.

But we know that some things topple over very easily, while others remain securely at rest. A die will lie steadily on whichever side it falls, but if we heap a number of dice upon each other, we shall soon erect such

---

* The word "*force*" is now, in strictness, used to denote the cause of motion, and "*energy*" to indicate the amount of work a force can do. To enter here, however, into any controversy as to the uses of such terms, would be foreign to the purpose of a work which proposes only to introduce students to the elements of science. We therefore do not hesitate to employ a popular phraseology when it seems likely to help on the purpose we have in view.

a pile of them as a very slight disturbance will cause to fall.

In very stable structures, like the ancient Egyptian buildings, such pressure was most amply provided for, and an equilibrium of the most stable kind produced.

This was the case above all with the Pyramids, and to a less degree in such temples as those of Philæ and Karnac —so impressive from the superfluous strength of their many rows of close-set, massive columns. In Grecian buildings we meet with the same secure repose, but in greater delicacy of build. In the arch and the dome, however, and still more in pointed architecture, the conflict of stones which tend to fall in different directions, produces (by the neutralising of each other's thrusts) a different kind of equilibrium and one of a less stable character.

FIG. 3.

As every one knows, substances of the same size of various kinds may be very different in weight; as we see in a cube made of cork and another of precisely the same size made of lead, or two glass vessels of the same size, one filled with water and the other with quicksilver or "mercury."

Heavier bodies are said to be more *dense* than lighter bodies of the same size. Thus experiment shows us that the *density* of Mercury is 13.6 times that of water. Similarly the *density* of lead is much greater than that of wood. Substances also differ in the extent to which they can be compressed or stretched or bent or twisted; as to the degree of elasticity they possess, or as to the ease with which they can be broken. They differ, besides, as to the amount of resistance (or *friction*) caused by the movement of one upon another.

In the elementary theoretical, as distinguished from practical, science of mechanics, however, the consideration of such differences is omitted in order to reduce problems to their simplest form. Solid bodies are supposed to be incapable of any change of form and perfectly inflexible, while cords are treated as perfectly supple and entirely devoid of any rigidity. Therefore the solids, fluids, and aëriform bodies of mechanics, are imaginary substances, and not such as we actually find in nature. But the consideration of these qualities and properties (of which abstraction is thus, for convenience, made) can always be added, and so the results of the science (as we saw* was the case with mathematics) correspond, with practical exactness, to the characters and properties of real material things.

Now every such thing may, for convenience, be supposed to be made up of an immense number of most minute and uniformly distributed particles; and the influence that makes it fall—which is known as the *force of gravity*, or *gravitation*—might be represented by lines drawn from every such particle towards the centre of the earth. But as such lines would thus converge towards a point enormously distant, they may be treated as if they were all parallel to one another.

But all such parallel forces (so represented by lines) may be replaced by a single force—also represented by a line—applied to a certain point, and such point is called

---

* See *ante*, p. 33.

"the *centre of gravity*" of the body—as being the centre of such parallel and equal forces—and is a fixed point which does not change, whatever be the position which the solid body may assume.

In order that a body should be in equilibrium, it is necessary for it to be supported by a force equal to the body's weight and acting through its centre of gravity in a direction opposite to its weight—as in the architectural illustrations just given.

In a cylindrical body, this centre is in the middle of its

Fig. 4.

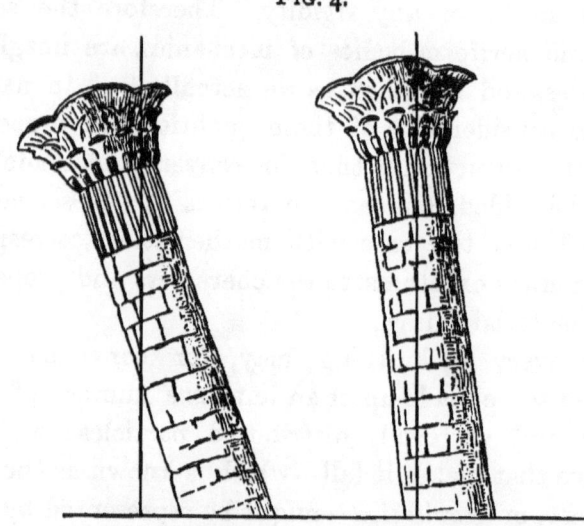

axis, and if such cylinder be obliquely placed (as in the Leaning Tower of Pisa) it will not fall, provided the weight at the centre of gravity be sustained—*i.e.*, if the vertical line from it to the ground comes within its basis of support. If it passes outside this, then such a cylinder, or building, must fall, and this is the reason why a very high pile of dice may so easily be made to topple over; because a very slight inclination will carry the centre of gravity of such a body beyond its base. Bodies, of course, may remain in equilibrium by being suspended—*i.e.*, by having the weight at their centres of gravity supported from above instead of from below.

If a body be supported in such a manner that on its equilibrium being disturbed, it tends to regain it (as in an oscillating pendulum or a detached wooden ball loaded with lead at one place) it is said to be in *stable* equilibrium. In the opposite case (as when a pole is balanced at one end), the equilibrium is unstable, because when disturbed, it tends to fall further away from, instead of regaining, the position in which the vertical line from its centre of gravity falls within its base.

The centre of gravity is not necessarily within the solid body itself which has to be supported, but may be in its vicinity, as in the case of a ring, or any hollow vessel. Thus it is that a variety of posturing tricks can be performed in tight-rope dancing. The dancer carries a long pole, the weight of which transfers the centre of gravity to the middle of the pole within the grasp of his hands so that he has it under his control. Similarly, in balancing rods on head or hand, the performer's art consists in keeping, by means of constant movement, the base of the rod under the centre of gravity. A number of ingenious toys are also constructed on the principle of an external position for the centre of gravity. Thus the figure of a prancing horse may be made to rock backwards and forwards, resting, near the edge of a table, on its hind feet only, in an apparently impossible position, by means of a leaden weight at the end of a curved wire, the other end of such wire being fixed to the belly of the horse, so that the centre of gravity of the whole structure is thrown behind and below the prancing figure. Thus, although its position looks most insecure, its equilibrium is really quite stable—namely, that of an ingeniously contrived oscillating pendulum.

A body is in equilibrium, or a state of rest, when the forces which act upon it counterbalance one another. One such force may be the force of resistance which a supporting body offers to the weight of the body which it supports and so keeps in equilibrium.

As numbers and quantities may be represented by arithmetical or algebraic symbols, so forces may be represented by lines of definite lengths. These will be the longer the greater the force they represent, and they will also serve to indicate the direction of the forces.

If two forces be in equilibrium, they must be equal in magnitude and opposite in direction. It is plain that if such were not the case, the greater of the forces would overcome the other, therefore the two would not neutralise each other, and so we should have motion, and not equilibrium.

But whatever the number and direction of the forces which may act upon any point, they can only produce motion in one direction. This is called the *resultant* of such forces, which are the several components of this resultant.

When two or more forces act on a point in the same direction, the resultant must be equal to their sum, and if in opposite directions, then to the difference between their sums.

Thus if any point be pressed upwards by a force of ten pounds and downwards by a force of five pounds, the resultant must be a pressure upwards of five pounds. If the pressure towards the west be $3 + 5 + 9$, while that to the east be $7 + 6 + 4$ the resultant $= 17 - 17$ or 0, which is equilibrium.

As to direction, let the equal forces A and B act

simultaneously on the point P, the force PC tending to draw the point P towards C, and the force PE tending to draw the point P towards E, then if from the two points C and E, equidistant from P, we draw two lines, CD parallel to PE and ED parallel to PC, meeting at D, the line PD will be the diagonal of the figure PCDE, and will represent the resultant of the two forces A and B.

The points C and E have been here made equidistant from P, because the two forces are supposed equal. Were they, however, unequal, then the distance PC

FIG. 5.

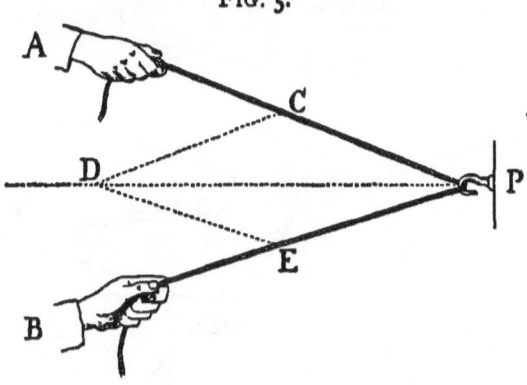

would have to be made to bear the same proportion to EP as the force A has to the force B—a unit of length representing a unit of force. In that case, instead of forming a parallelogram (*i.e.*, a straight-lined quadrilateral figure whose opposite sides are equal and parallel) with adjacent sides equal, such sides would be unequal. Now, it is a rule in mechanics, that if two forces acting at a point be represented in magnitude and direction by the sides of a parallelogram, the resultant force will be represented in magnitude and direction by the diagonal of the parallelogram passing through that point.

Any number of forces acting on a single point can be

computed by this rule, and, inversely, any single force may be considered as resolved into any number of forces of which such single force would be the resultant. The former process is called the *composition* and the latter the *resolution* of forces, and both processes are most frequently employed in the science of mechanics.

In a system of balanced or *statical* forces, each is exactly equal and opposite to the resultant of all the rest, as is shown in a proposition known as the *Polygon of Forces*.

Let us suppose that five forces are all acting at (from)

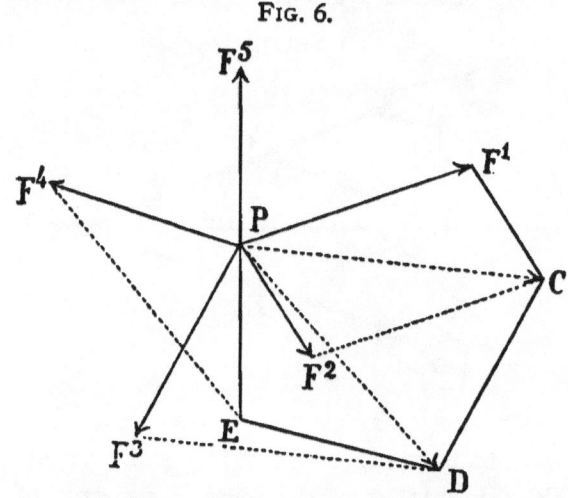

FIG. 6.

the point P, while their respective directions and intensities, are represented by the five lines PF^1, PF^2, PF^3, PF^4, and PF^5, passing from those letters to P. Constructing a parallelogram whereof PF^1 and PF^2 are two sides, the line PC will be their diagonal and resultant. Next taking this resultant and the next force PF^3 as two sides of another parallelogram, we find that PD will be their resultant, and therefore the resultant of all the three forces. Finally taking this latter resultant and constructing a parallelogram from it and the fourth force, we find that PE is the resultant, which we see exactly balances the remaining force PF^5. Thus is formed a many-sided figure or polygon, consisting of the lines PF^1, F^1C, CD, DE, and EP, the five sides of which represent the five forces, because a parallelogram must have its opposite sides equal.

PF^1 expresses the intensity of the force acting along PF^1.

F^1C expresses the intensity of F^2, because F^1C is the side of a parallelogram whereof PF^2 (expressing the intensity of F^2) is the opposite side.

Similarly CD must equal PF^3 and DE must equal PF^4, while $PE = F^5$.

There may be two forces acting side by side, as in a two-horsed carriage. Two such powers are called *parallel forces*. The resultant of two parallel forces acting in the same direction is equal to their sum, and, when such forces are equal, the resultant of their combined force acts midway between the points of application of each; when unequal, it is in a definite degree (as we shall shortly see) nearer to the stronger force.

When the parallel forces are equal but act in opposite directions, their result is to produce rotation, and this tendency cannot be counterbalanced by any single force.

A practical knowledge of rudimentary mechanics was no doubt early obtained, since human ingenuity would readily suggest the application of a strong stick, as a lever, to raise a heavy body from the ground, and would lead to the perception that it is easier to push such a body up a sloping surface, than to raise it in men's arms and carry it. Though we can introduce the reader but to the first elements of mechanics, we must nevertheless offer some explanation with respect to the principle of the lever, the inclined plane, and the pulley, referring him for further explanation about them to professed treatises on mechanics.

To return for a little to the consideration of the action of two parallel and *unequal* forces:

Let us suppose that a heavy rigid bar is balanced on a point at its centre, which must, of course, be its centre of gravity.

Now if we suppose the bar to be made up of two bars, a larger one, AD, and a shorter one, DB, then they also can be supported at their respective centres—namely, at D′ for AD, and at D″ for DB. But the two bars thus respectively supported at D′ and D″, act as two parallel and unequal forces (namely, the weight of each), and their resultant must pass through the point

FIG. 7.

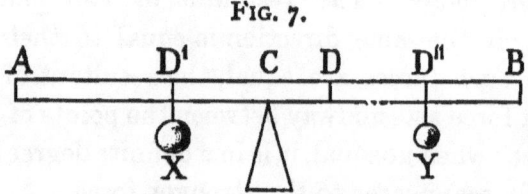

C, because it is at that point that their two pressures are neutralised by the support which balances the whole. Hence we see that the resultant of the two unequal forces does not pass through a point midway between them (*i.e.*, mid-way between D′ and D″), but through the point C, which is much nearer to D′ than to D″. It is just so much nearer as the weight of AD is greater than the weight of BD. That is to say, as D′C : D″C :: the weight of BD : the weight of AD. In other words, the resultant of the two unequal parallel forces is so situated that its distance from either, shall be inversely as their intensities. To prove the truth in the example chosen, we may suspend from the points D' and D" two additional weights X and Y, bearing the same ratio to each other as the weights of AD and DB had previously borne. Then the balance will still remain undisturbed in spite of the greater weight suspended on one side of C compared to that on the other side of it.

A lever is a rod we will suppose to be perfectly rigid

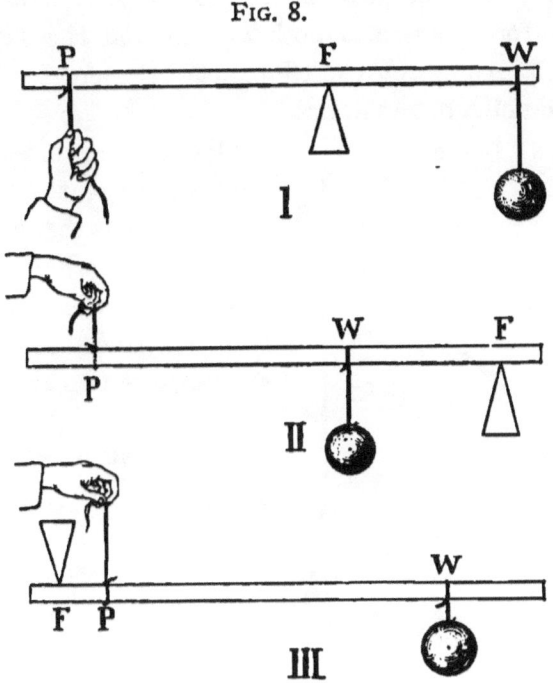

Fig. 8.

(and we will here assume it to be also straight), which turns on a fixed point called the *fulcrum*. A force is applied at some point in the lever, while at some other point there is a resistance acting, which resistance the force has to overcome. The portions of the lever which may be on either side of the fulcrum are called the *arms* of the lever.

Levers may be of three orders: a lever of the first

order (I. Fig. 8) is one where the force or effort acts on one side of the fulcrum, while the resistance is on the other.

A lever of the second order is one where both forces are on the same side of the fulcrum, while the point of resistance is placed between the fulcrum and the effort. (II. Fig. 8).

A lever of the third order (III. Fig. 8) is one where the two forces are also both on one and the same side of the fulcrum, but the effort acts between the fulcrum and the point of resistance.

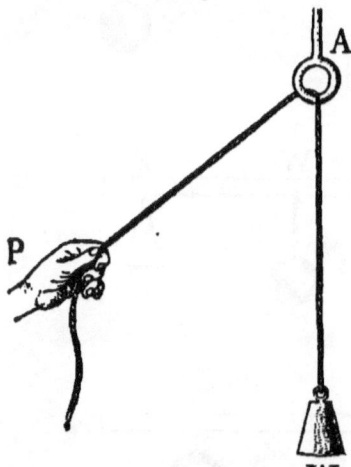

FIG. 9.

One of the most useful applications of the principle of the lever is that employed in the *balance,* which is a lever of the first order, called the *beam,* suspended at its centre and with its two arms consequently equal.

In the *pulley,* we have an example of the transmission of force through a cord considered as perfectly flexible —free from all friction, and inextensible.

Thus the resistance of a weight suspended at one end of a cord, may be perfectly balanced by a force (such as a sufficient grasp of a hand) applied at the other end of a cord which has been passed over a hook and through a ring (A, Fig. 9).

Such forces must always be what is called *divergent,* or opposed to each other, and by such an arrangement the direction in which a force acts may be changed without modifying its intensity, the cord being supposed always to undergo the same tension at every part of its length. It is practically convenient to use a pulley instead of a hook or ring; the pulley being a wheel which turns freely on an axle passing through its centre, with a groove at its circumference for the reception of the cord. Nevertheless it is the cord and not the pulley which is the efficient agent in this mechanical contrivance. By various ingenious arrangements of pulleys, a very small force may be made to overcome a great resistance, as for instance in lifting a heavy weight. But for a description of these the reader is referred to professed treatises on mechanics.

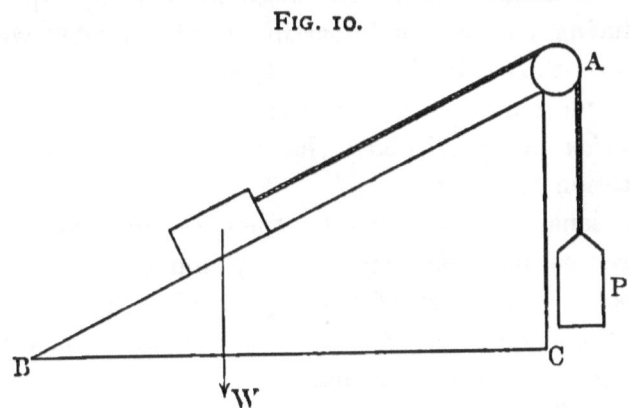

FIG. 10.

The *inclined plane* is another contrivance, by means of which a weight may be lifted to a certain height by the application of a force less than itself. The extent of the inclined surface, AB (Fig. 10), is termed the length of the inclined plane. Its height is represented by AC, and its base by BC.

Suppose a heavy weight W to be supported on a smooth inclined plane ABC by a force (weight) P, as in the figure—*i.e.*, by a cord passing over a pulley at A. Then it will be seen that if P exactly balances W, and there is no friction, and if the body W ascends from B to A—*i.e.*, if it *rises* through the distance CA—the weight P *descends* through a distance equal to BA; and by comparing P with W it will be found that

As P : W : : AC : AB.

Consequently, by diminishing the *height* of the plane, as compared with the length, the force P may be made as small as we please, compared with W, the resistance to be overcome.

We have hitherto considered force with respect rather to its statical effects as equilibrating other forces; but force is manifested in its most interesting aspect in producing motion—in imparting motion to what we call matter—the study of which is dynamics.

We must now consider a little more fully the subject of *motion*, and particularly that of bodies moving under the action of forces.

Motions may vary as to their *velocity*, and these differences may be expressed by numerical or other symbols. The idea of velocity involves the ideas of *time* and *length* and *direction*—as we speak of a body moving in a given direction at the rate of "ten miles an hour," and the unit of velocity can be defined only by reference to the unit of length and the unit of time.

"Velocity" is sometimes called the "intensity of motion," but the *quantity* of the motion of any body, which is called its "momentum," is very different from its velocity, and refers jointly to the amount of matter of which a moving body consists as well as to its velocity. Thus if two bodies are moving with equal velocity, the quantity of motion in each corresponds with its mass, and if two bodies of equal mass are in motion the quantity of motion in each will then be proportional to its velocity.

If two unequal bodies are moving with different velocities, their quantity of motion is jointly proportional to their respective masses and velocities.

But motions may not be of uniform velocity during the time they last; they may be continually *accelerated* or *retarded*, so that their velocity varies from moment to moment owing to some accelerating or retarding cause.

Matter itself must be regarded as absolutely inert—not inert, however, in the sense that it is more inclined to rest than to motion, or that motion naturally tends to come to an end. By calling motion "inert," it is simply meant that matter is totally indifferent to either rest or motion, and therefore it has been purposed to speak of this quality as *persistence* rather than *inertia*.

The following are Newton's three laws of motion:

(1) *Every body continues in its state of rest, or of motion in a straight line, except in so far as it is compelled by impressed forces to change that state.*

Now there is no such thing as absolute rest, since, as we have seen, every body tends to move in the direction in which gravity draws it, and only does not so move because some other force prevents it. Therefore what the first part of this law really asserts is, that when a body is maintained in a certain state and position by the combined action of two or more forces, such state will continue till some other force changes the conditions.

The second part of the law affirms that a body in motion tends to move uniformly in a straight line. This necessarily means that its movement must continue in the same direction, until some other force changes that direction.

(2) *Change of motion is proportional to the impressed force, and takes place along the straight line in which that force is impressed.*

This means that whatever motion (and by motion is here meant quantity of motion) any force produces, twice or three times such force, or such force acting for twice or three times the duration, will produce twice or three times as much motion, and so on.

Therefore when several forces act together, the change of action due to each is proportional to each, and their combined effect must be the same as if each had acted separately or successively. Any body simultaneously acted on by two or more forces will be carried to the spot it would eventually have reached had the same forces acted separately and successively.

(3) *To every action there is an equal and opposite reaction; as the mutual actions of two bodies on one another are always equal and in opposite directions.*

In other words: any body set in motion by another body will react upon the latter in an opposite direction, and the second body will lose a quantity of motion exactly equal to that which the first received.

Thus if any body A, exerts a force on another body B, B must also exert on A an equal force in an opposite direction.

Thus every force is, in fact, one of a pair of forces, and such a pair of forces is called a *stress*. We have an example of a pair of forces of the kind in those which lead a body revolving in a circle, respectively to approach and to fly away from that circle's centre. These two forces are respectively known as "*centrifugal*" and "*centripetal*" forces.

MECHANICS

Here, as before, when dealing with geometry,* what we see in real life does not exactly correspond with abstract scientific principles. All the motions we observe about us sooner or later come to an end, and no body propelled from the earth's surface goes on long in one direction, but sooner or later descends to the earth. These facts are, of course, due to friction which, in different degrees, retards motion, and to the force of gravity which draws all things that may be propelled from the earth's surface downwards again towards its centre. Therefore in many dynamical problems we have to neglect the consideration of friction. But friction may not only be more or less diminished, it may be actually neutralised by the action of some other force.

Thus a railway train once set in motion would, according to our first law of motion, continue onwards uniformly if its motions were not retarded by any other force; but friction tends to prevent this, and it would soon stop the train but that the force generated by the engine is sufficient to overcome the impeding influence of friction. The train will thus continue onwards in uniform motion under the influence of opposing forces.

The motions which pertain to any separate body, continue unaffected by a motion common to it and other bodies also—*e.g.*, a watch will continue its proper movements while in the pocket of a man running a race. This truth is connected with the second law of motion, which affirms the effective independent action of forces apparently combined.

It may be illustrated by the fact that a weight

---

* See *ante*, p. 36.

dropped from the top of the mast of a ship in rapid motion will fall on the same spot as it would do were the ship at anchor. For it participates in the onward motion of the ship, and this horizontal impulse prevents its being left behind by the motion of the ship during the time of its descent.

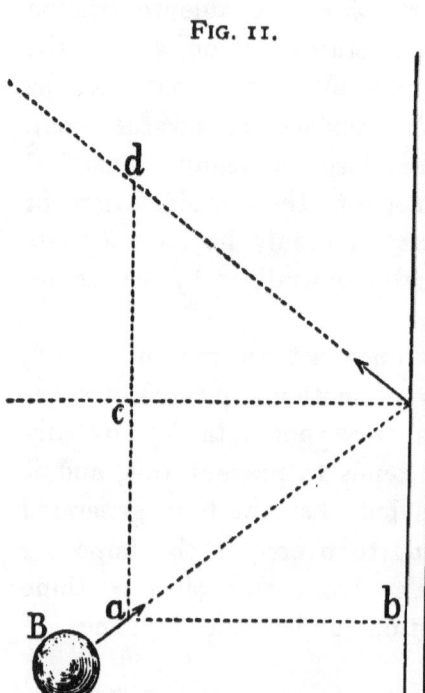

Fig. 11.

It may also be illustrated by the impulse given by a billiard cue to a ball B, by causing it to strike against the cushion of a billiard table at the point X (Fig. 11). As we have seen, this force, represented by the line from a to X, may be resolved into two forces represented by $a\,b$ and $a\,c$. Their combined action (represented by the diagonal aX) would bring the ball to the point X. There the force Xc would, by the third law of motion, cause a reaction by the cushion on the ball, tending to drive it back along the parallel line Xc. For only the force ab has acted on the cushion, while the force ac has met with no resistance. This last force, then, is still in full operation, and acting together on the impulse Xc, carries the ball to the position d. On comparing the diagonal BX with the diagonal Xd we see that the angle BXc equals the angle cXd, or, in other words, that "the angle of reflexion equals the angle of incidence"—a truth of the greatest value to billiard players though they have to allow for the friction and other conditions which prevent this equality being attained on any billiard table with absolute exactness.

When a body is once in motion, force is not needed to maintain the motion. When, however, there is any change in the direction or speed of a moving body, then we have evidence of the existence of force. This is only another way of stating the first law of motion.

Therefore any continued force must produce a continuous change, either in direction or velocity. A sudden change of either kind is produced by *impact*, *i.e.*, by an instantaneous exertion of force.

From what we saw with respect to *quantity** of motion, it may be approximately deduced that a charge of gunpowder which would impart to a bullet of a certain size a velocity which we may express by 100, would impart to a bullet ten times that size only a velocity of 10.

Various interesting apparatuses have been invented to illustrate these laws of motion, but for their description and a vast mass of further information, the reader must have recourse to distinct treatises on dynamics.

When a force acts continuously upon a body, the effect is necessarily cumulative, and in that case its velocity will be constantly quickened and *accelerated*. If we let fall from our hands at the same time a feather and a marble, the latter falls at once very quickly to the ground, while the other falls very slowly and with many oscillations.

---

* See *ante*, p. 54.

The reader will at once understand that it is only the resistance of the air which prolongs the feather's descent. Accordingly, if they are made to fall in the nearest approach to a vacuum which an air-pump can produce, they will fall simultaneously.

If a bullet be taken in the right hand and be allowed to fall thence into the left hand through a height of a few inches, it will give a slight blow. If it be allowed to fall a yard, it will be felt more smartly, while if it were to descend on the hand from a second-floor window the blow would be severe.

Therefore the longer a fall lasts—the greater the distance through which a body falls—the greater is its energy and the greater the rapidity of its motion. The motion of a falling body is a *uniformly accelerated* motion, because, the attraction between the earth and the body never ceasing to act, the body gains a fresh momentum every instant. Therefore it falls to the ground with a velocity which is the aggregate of all the indefinitely small but equal increments of velocity thus communicated to it.

Now it has been discovered that a falling body acquires, at the end of the first second of its fall, a velocity of about 32.2 feet a second—*i.e.*, a velocity which would, alone, carry it through, say, 32 feet during the second second. During this second, however, it will have fallen through only 16 feet. During the second second it will fall through the 32 feet (from the velocity with which it starts) and through 16 additional feet on account of the constant action of the force of gravity. Similarly, it will fall 64 feet during the third second plus 16 feet, and so on, as shown by the following table wherein the time of each second is represented by a similar length, velocity by breadth, and the distance fallen by extent of area:

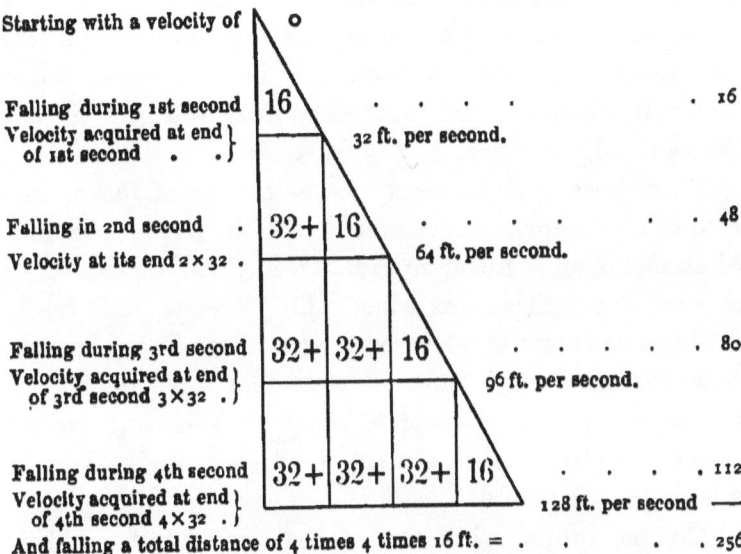

FIG. 12.

But though a body falls 16 feet in a second, it will only fall 4 feet in half a second, for it is then falling at only half its speed per second. Thus, as Galileo observed, the distance fallen is proportional to the square of the time occupied by it.

Thus if we know the time which has been occupied by a fall, we can determine the space through which it has fallen by multiplying the square of the time by the number of feet through which a body falls in one second.

During one second it will have fallen 1 × 16 ft. = 16 ft.
 „ two seconds „ „ „ „ 4 × 16 „ = 64 „
 „ three „ „ „ „ „ 9 × 16 „ = 144 „
 „ four „ „ „ „ „ 16 × 16 „ = 256 „

It also follows from the foregoing facts that the velocity a body acquires in falling, is as the square root of the height fallen through. Thus, to acquire a velocity of 32 feet per second it must fall a distance of 4^2 feet; for a velocity of 64 feet per second, 48 feet; and for 96 feet per second, 80 feet and so on. The distances fallen through during equal successive intervals are as the series of odd numbers 1, 3, 5, 7, &c.

These laws apply not only to the motion of *falling* but also to all uniformly accelerated motions; only the rate of acceleration is never so rapid on inclined planes or in any other conditions, as when falling freely. All freely falling bodies are accelerated at the same rate, because however they may differ in mass, the force of gravity acts on them in exact proportion thereto—acting twelve times more forcibly on a mass of twelve pounds, than on a mass of one pound.

The pendulum, while one of the simplest of scientific instruments, is also one of the most valuable. If a small, heavy body be suspended by a thread from a fixed point, that will form an instrument of the kind of a most simple description.

When at rest, the line from the point of suspension S (Fig. 13), to the weight A, serves to indicate the line along which gravity acts—the "plumb-line" or vertical line.

When the weight is drawn on one side, and then let go—*e.g.*, if A be drawn to C and then allowed to fall in a vertical plane—it will, after descending to its former position, ascend on the other side as far as B— that is, nearly as far from A on one side as C was on the other. It will then descend again and afterwards ascend nearly as high as was the point B, and so on. Its entire sweep from C to B is called one *vibration*, or oscillation of the pendulum, and its extent or *ampli-*

MECHANICS

tude is measured by degrees, minutes, and seconds of an arc which may be placed so as to measure it. 360 degrees have been adopted as subdivisions of a circle, each such "degree" being subdivided into 60 minutes and each such minute into 60 seconds.

The time occupied by a pendulum in one oscillation constitutes the *duration* of a vibration. Were it not for friction and the resistance of the air, the weight would ascend always to the same height on either side as that whence it first started, and so would constitute an instrument with perpetual motion, since the action of the force of gravity is incessant. Within certain limits, the time occupied by a vibration is not altered by increasing its amplitude, because the more the weight be elevated, the more the speed of its descent will be increased, and in exact proportion to the degree of elevation.

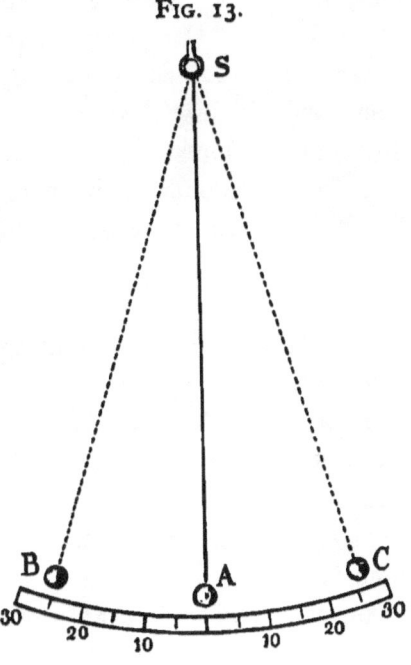

Fig. 13.

The vibration of pendulums being thus a simple and direct effect of the force of gravity, they have been made use of to measure variations in that force at different places, to estimate the density of matter beneath the surface of the ground, and even to determine the shape of the earth.

It has been ascertained that the time occupied by a

pendulum in its oscillation, varies as the square root of its length, thus four pendulums, the relative lengths of which may be represented by the numbers 1, 4, 9, and 16, will oscillate in periods represented by 1, 2, 3, and 4.

All that has been said with respect to uniformly *accelerated* motion applies equally to uniformly *retarded* motion. Thus when any body is projected straight upwards from the earth's surface, it rises 32 feet less during each succeeding second, till its velocity (which is decreased during the ascent as it increases during a descent) is exhausted. Thus it must pass each successive point as it descends again, with the same velocity as that it possessed as it passed each such point during its ascent.

But we have constantly to consider the joint effects of a body with uniform motion and a *uniformly accelerated* motion—as, for example, when a shot is fired from a cannon. Such a body is impressed with the uniform motion imparted by the explosion and with the uniformly accelerated motion due to the force of gravity. Putting entirely aside the action of friction and atmospheric resistance, we find that there is an exact composition of forces. Thus at any moment the cannon ball will be at the spot it would have reached had it been carried, in a straight line, to the elevation it would have attained by the force of projection acting alone, during the time elapsed, and then fallen thence in an exactly similar time. The junction of all these points of coincidence—*i.e.*, the path followed by the projectile always forms a peculiar curved line called a *parabola*—a curve such as would be produced by the margin of a section of a circular cone cut through parallel to any part of its slanting surface.

Another, and a most noteworthy, conic section* is one formed by cutting across a cone in any place, not at right angles to its base. It may be drawn in the following simple way: If two ends of a thread be attached to two points of a horizontal surface—the thread being much longer than the distance between such two points—and if a pencil be so placed as to stretch the thread outwards as much as possible, and then be carried round (always so stretching the thread) till it describes a closed curve, such a figure will be an *ellipse*. The two points of attachment of the thread, form what are called the *foci* of the ellipse, and the more these are approximated the more circular the ellipse will become; and it becomes transformed into a circle as soon as they coincide.

Now if a body were projected horizontally from a point external to the earth's surface with sufficient velocity, it would be carried in a certain time to a much greater distance than gravity would make it fall during that same time, and then (if there were no air) it would never fall to that surface, but would continually go round the earth in an ellipse—the precise form of which would depend on the exact velocity and direction given to the body.

Every stone flung into the air describes a little bit of an ellipse round the centre of the earth, which it would complete but for the overpowering attraction of the earth.

It has been ascertained that such a body, whatever the amount of its divergence from a circle—*i.e.*, whatever the eccentricity of the ellipse in which it migh. move—would be subject to the following law:

---

* See *ante*, p. 37.

A line drawn from it to the centre of the earth, must always move in the same plane, and in such a way as to pass over equal *areas* in equal *times*. Such a line is called a "*radius vector*."

But the force of gravity between bodies does not alone draw everything at or near the earth's surface, towards the earth's centre, it also draws every existing body towards every other, although its action between small bodies is too feeble to be easily observed.

Each body thus draws towards itself every other body with a force of gravity which varies directly as its mass and inversely as the square of its distance from the body it attracts.

On account of the inertia of matter, or its absolute indifference to motion, every separate body on the earth's surface would, by the force of the earth's rotation, be projected and continue onwards in a straight line from its surface—in a *tangent**—were it not for the force of gravity which keeps it in its place. The passive tendency to continue onwards in a tangential straight line—or "to fly off"—is, as before said, termed "centrifugal force," while the action of gravity which conflicts therewith is called "centripetal force." These two forces arise together, and illustrate that bifold nature of forces implied in Newton's third law.†

Now, as we all know, the earth revolves on its axis once in every twenty-four hours.

The weight of any object, then, is that portion of its gravity over and above that which is required to prevent its "flying off" (owing to our globe's rotation), and to retain it on the earth's surface. Had bodies no more

---

* A tangent is a line touching the circumference of a circle and at right angles to the diameter of the circle at the point of contact. † See *ante*, p. 56.

gravity than would be required to effect this, they would have absolutely no weight and would exercise no pressure whatever. It is the tendency to "fly off" from a horse's back produced by rapidly riding in a circle, which so reduces the weight of a circus rider that he can easily stand on the saddle and perform a number of feats any one of which would be impossible did he ride in a straight line.

On account of the greater rapidity of motion of the earth's surface towards the equator than towards the poles, the centrifugal force is necessarily greater at the equator, and consequently weight is there slightly diminished, as is easily proved by the vibrations of a pendulum. On account also of the greater mass of the earth's equatorial region, a plumb line does not, in the north, hang absolutely vertical to the earth's surface, but deviates slightly to the south.

With these various elementary observations we must conclude what we have to say respecting the mechanics of solid bodies, referring the student to other works for the prosecution of his study of that science.

Passing on now to the consideration of fluid substances, we may first remark that the essential principles of dynamics, apply to them as well as to solids, but the fluid condition calls forth new conceptions, which are treated of as distinct sciences known as *hydrostatics*, *hydrodynamics*, and *pneumatics*.

All fluids, whether liquid or aëriform, are, of course, no less subject to the action of gravity than are solids, but the commonest observation makes it clear that the internal constitution of their substance must somehow be very different from that of solids. In what precisely that difference consists we do not know, though any speculations are useful, provided that (as working hypotheses) they serve to help us on to a better knowledge of the laws which govern fluid bodies.

As to liquids, we may assume them to be made up of particles, which, instead of cohering stably in some definite order (as we assume to be the case with the particles of solid bodies), have no tendency to preserve any reciprocal positions, but can move and glide over each other with perfect freedom and in all directions, each particle pressing equally on all the particles which surround it and being equally pressed on by them.

By this hypothesis we may understand the great difference which exists between a liquid and a solid. Hitherto our conception of *pressure* due to *weight*, has been simply downwards, through the force of gravity. But a portion of liquid presses equally in all directions, in consequence of the action of gravity, or of any other force acting upon it.

Therefore for liquid to be in equilibrium, every particle of it must press and be pressed upon equally in all directions. One consequence of this is that the surface of a liquid, apart from any disturbing influence, must be horizontal. An illustration of the imagined condition of liquids, may be obtained by considering the consequences which would be produced should a fresh comer try to effect an entrance into a room already filled by a crowd of persons. The new comer who manages to effect such an entrance will produce pressure in all directions—on every side of him.

From the mobility of its particles it follows that a liquid immediately takes the figure of any vessel in which it may be received. Therefore, if two or more vessels, however different their sizes, which contain liquid of the same kind, be placed in communication (*e.g.*, by turning stop-cocks) below the surface of the liquid in any one of them the fluid in the whole of them will settle itself at one and the same level, or, as is commonly said, "water will always find its own level."

It has been said that liquids transmit pressure, not only in the direction opposite to that in which it is applied, but equally in all directions; thus a hollow vessel may be filled with water, which it will hold quite securely. But if a long tube be screwed into the top of it and filled with water, it will cause the so-filled vessel to burst, if the tube and column of water be sufficiently

FIG. 14.

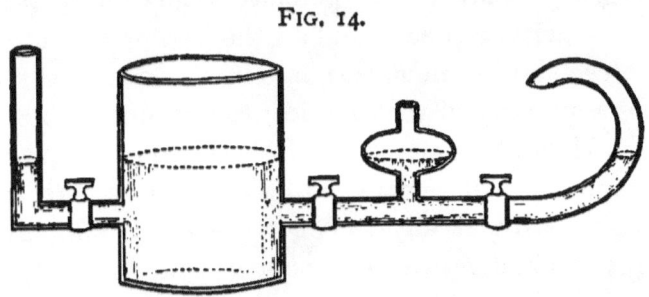

high; and it makes no difference whether the tube be stout or slender.

Therefore the weight of even half an ounce of water will burst any vessel, if the tube and column of water are only high enough; for on account of the equality of pressure in all directions, a pressure equal to the whole weight of the entire column will be exerted on every part of the inner surface of the hollow vessel, which is of equal size with the bore of the elongated tube.

In hydrostatics, it is assumed that liquid bodies are practically incompressible. Such is not actually the case, though pressure will only reduce their bulk so inconsiderably that the result of such action may practically be disregarded.

The pressure of water of the height of one foot is

about half a pound for the square inch; and as we increase the height a foot, the pressure increases half a pound. In a cubical vessel the pressure of a liquid filling it is, as before said, equal on all sides, and its pressure on each side is equal to half the weight of the liquid. Therefore a liquid in a cube exercises, on base and sides, three times as much pressure as that produced by its weight alone. Let us suppose its weight exercises a pressure of one pound, then the pressure exercised on each side of the cube will be half a pound—that is, a pressure of two pounds, besides the pressure, due to gravity, of one pound on the base of the cube.

Any solid body immersed in a liquid, necessarily displaces a quantity of that liquid exactly equal to its own bulk. If it also exactly equals this displaced quantity in weight, it will remain indifferently at any depth in the liquid without any tendency to rise or sink. If its weight is greater it will of course sink, and if less, it will rise. Not that, of course, it has any spontaneous tendency in itself to rise; it simply rises because the greater pressure pushes it upwards. But a body which sinks, apparently loses just as much of its own weight as the water it displaces weighs, as may easily be ascertained experimentally.

Since liquids press equally in all directions, any object immersed in them must be at least as much pressed upwards by pressure from below, as it is depressed by pressure from above. Thus fishes can swim with ease at depths where they must be subjected to enormous pressure from above, since they are sustained by a somewhat greater pressure from below.

If a solid body be first weighed in air and then in water; if its weight in the latter be subtracted from its weight in the former and its weight in air be divided by the difference, the product will be what is called the *specific gravity* of that solid. Let us suppose a solid weighs 75 grains in water, and 80 in air; then $80 - 75 = 5$ and $\frac{80}{5} = 16$. The proportion, therefore, of the weights of equal bulks of the solid in question and of water, will be 80 to 5 or 16 to 1, so that it will be 16 times heavier than its own bulk of water, and the specific gravity of that body will be 16.

In England it is customary, for convenience, to consider one cubic foot as the standard volume, and to express the weight in avoirdupois ounces or grains, and a cubic foot of rain-water weighs about 1,000 oz. The specific gravities of liquids may be ascertained by using a vessel capable of holding 1000 grains of water at a temperature of 60° Fahrenheit.

If two liquids which differ in density—such as water and alcohol, or water and mercury—be made to communicate in a vessel with two upward prolongations, or limbs, then the height to which they will rise in the two limbs will differ inversely as their densities—the more dense being the less high.* Thus two inches of mercury will balance 27 inches of water.

We have said † that the surface of a liquid is horizontal, "apart from any disturbing influence"; but certain attractions, other than the earth's gravity, may interfere, more or less, with the horizontality of a fluid's surface.

Thus there is, of course, an attraction between the vessel holding a liquid and the adjacent portion of such liquid. If the solid vessel be denser than the liquid

---

* As to "osmosis," another effect of placing in proximity, see *post*, p. 146. † See *ante*, p. 68.

(as in the case of water in a glass vessel) then the surface of the liquid will rise at its circumference so that its upper surface is concave. If the liquid be much denser than the vessel containing it (as with mercury contained in glass), then the surface of the liquid will be slightly depressed at its circumference and the upper surface of the mercury will therefore be convex.

This attraction, which is due to the surface contact of the liquid and solid, is shown on plunging any solid body into water, when some of the water adheres to it and comes out wet. If thin plates of different substances be made to touch the surface of water, considerable force is required to raise them from it, and the amount of force thus required varies in amount with the nature of the substance employed. One conspicuous form of such action is that which is known as *capillary attraction*. This is the attraction exercised by tubes of very fine bore upon liquids into which their ends may be plunged.

If any substance containing minute canals of the kind, be immersed in water, then the water will ascend them to a height which will be the greater the narrower the cavities it ascends. As examples of such action may be taken the small cavities in blotting paper, sponge, the cotton of a lamp, lump sugar, &c., in all of which liquid will readily ascend. It is because such cavities are generally not broader than a human hair that this attraction has been termed "capillary."

The motions of liquids constitute another section of mechanics known as hydrodynamics or *hydraulics*. Since liquids can be set in motion with so much greater facility than solids, and since the direction and velocity of their movements are liable to modification by so many causes which would not modify the action of moving solids, it is evident that the conditions of their movements must be relatively complex. Nevertheless the motions of liquids have the same basis and obey the fundamental laws of the movements of solid bodies.

We have seen that, abstracting the action of the atmosphere, all bodies which fall from the same height fall simultaneously and attain the same velocity at each stage of their descent. We have also seen* that the velocities they require are as the square roots of the heights through which they fall, so that an object must be four times higher than another if we desire that it should attain twice the velocity of the latter.

Let us suppose that we have two vessels before us, each containing a depth of four feet of water, but that one vessel is six feet in diameter, while the other has but a breadth of one foot. Further let us suppose that a similarly sized hole be made in each vessel six inches from the bottom. It might be thought that the stream issuing forth from the larger vessel will be projected much further than that from the smaller one. Such, however, is not the fact; they will be projected equally far. Not only is this the case, but it will be the same if the two liquids are of different densities. Mercury will be projected as far, and no further than, water, if they both issue from similar orifices, placed at the same depth beneath the surfaces of the two liquids. Here, as with solids, if we wish to double the velocity we must raise the surface of the fluid fourfold, and to make it four times as great we must raise it sixteenfold and so on; and the greater the velocity, the further outwards will the jet of discharge extend. This jet, in falling, always describes a parabola, *i.e.*, falls in a parabolic curve. The particles

---

* See *ante*, p. 60.

of water, in issuing from an orifice of a vessel (at the bottom), do not, as Newton has shown, pass perpendicularly to and through it. Many of them converge towards it from every side; so that after passing out of the orifice they form a stream of diminished breadth, which he called the *Vena contracta*. As the liquid issues forth, there forms on its surface (immediately over the orifice) a hollow depression which deepens till it forms a conical space, the apex of which is at the orifice towards which the liquid flows, while a rotary movement can be very easily given to, or transmitted through, its particles. This movement must also augment in velocity as the liquid escapes (and so diminishes the extent of the circling waves) through the inertia of the particles, which tend to preserve whatever velocity they may have gained, and therefore, as they approach towards the centre of the rotating mass, their speed must increase as the circles they form become smaller. If the orifice be closed, then the conical depression will wander over the surface of the liquid, gradually becoming shallower and shallower, till it disappears. The quantities which thus flow out, in successive, equal intervals, from the bottom of a vessel with vertical sides, are as the diminishing series of odd numbers 9, 7, 5, 3, 1— which correspond *inversely* with the spaces described in equal intervals by a falling body.*

The passage of liquids through pipes is greatly retarded by friction within, and by the resistance experienced when bends take place. The retardation may be diminished, however, by giving particular forms to the commencement and termination of a pipe without otherwise changing its capacity. Thus a 4-inch pipe

---

* See *ante*, pp. 61 and 62.

may (whatever its length) be made to deliver considerably more water if its first three inches and the last yard of its length, be enlarged and given a conical shape.

The motion of water in the bed of a river would, but for the resistance offered by its sides and its bed, go on continually accelerating from its source to its mouth, like a solid body falling by gravity. In that case enormous destruction would be produced in the lower lands while the upper parts would be deprived of all moisture. But the adherence of the particles of water together, and the friction against the sides and bed of the river, produce a resistance which increases with the velocity of the current, till it equals the accelerative force of the descent, and so a uniform motion becomes established.

Irregularities in the sides and beds of rivers often produce currents setting obliquely, or eddies, and of course the steeper the descent of the river's bed the greater the velocity and force of the current.

When a liquid has any part of its surface raised above, or depressed beneath, the rest, it will, as has been said, return to the general level. But in so doing it, like a pendulum, acquires a velocity which carries it beyond the position of equilibrium, and thus it oscillates, communicating similar oscillatory motions to the adjoining portion of the surface of the liquid, and that to the next and so on. But as all these communications of motion are not simultaneous but successive, an appearance is produced of an elevation, or of a series of elevations, travelling along the surface—in other words, we have what is called a *wave* or a series of *waves*.

Each wave contains particles of the liquid in all degrees of oscillation, elevation and depression, and the breadth of a wave is measured between particles which are in similar position, *e.g.*, from those at the greatest depression in front of and behind the wave. This dimension, like the length of a pendulum, varies as the square of the time of oscillation and the velocity of a wave varies as the square root of its breadth. Thus if a boat be noticed on one day to rise and fall twice as often as it did on the previous day, then the waves which pass under it must have become four times as broad, while moving with only double velocity.

The laws of hydrodynamics have led to the construction of aquatic machines for raising water. It would be quite beyond the purpose of this work to describe such in detail, but we must briefly refer to the screw of Archimedes, water-rams, and water-wheels.

The first may be either a flexible tube open at both ends and wound spirally on the exterior surface of a cylinder, or it may be a plate of metal coiled spirally about an axis enclosed within a hollow cylinder. The machine is fixed in an inclined position, with its lower extremity immersed in the water which is to be raised.

While it is at rest, the water occupies the lower part between two of the bends of the spiral. When turned, the machine is rotated on its axis, and the part containing the liquid being thus elevated, the water will be caused, by gravity, to descend into the lower part between the next bends of the spiral, and so, in reality, it rises with respect to its former position in the rotating spiral coil within which it is confined. Thus the water continually proceeds towards the upper part of the machine from whence it is discharged.

A water-ram is a machine by which the action of gravity on falling water is utilised by a succession of valves which hinder the moving water from returning while allowing it to pass freely in the opposite direction; by this means it can be raised to a much greater height than that from which it falls. It is so arranged that a stream of water is made, by its descent, to open and close a valve, which, each time it shuts, drives a portion of the water up another tube and to a higher level, where it is again retained by a valve, and so on.

In water-wheels, the liquid acts as a moving power by its weight, its momentum, or by both of them combined, acting on such wheels.

In the first case, a wheel is provided at its circumference with troughs into which the water is received near the level of the axle of the wheel; the vessels thus filled becoming heavier than those on the other side, the wheel is made to revolve by mere excess of weight.

In the second case, the water may fall into the troughs from a more or less considerable height above the axle, so as to add the increased effect of the momentum gained by it in its fall. This is called an *overshot* wheel. An *undershot* one has flat projections from its circumference, while its lower portion is plunged in a stream capable of turning it.

We must now pass to the last subdivision of mechanics, namely, that which relates to aëriform fluids, and which is known as the science of *pneumatics*.

Aëriform bodies differ greatly in nature from solids and liquids. In solids, the particles of which we may conveniently suppose them to consist, are stably held together or, as we have seen, cohere in varying degrees of tenacity. In liquids, the particles still cohere, but so unstably that they glide over each other with the greatest ease, and a liquid presses equally in all directions. In aëriform bodies, however, not only do the particles not cohere, but they actually repel each other and separate as far as possible, pressing, however, equally in all directions. This tendency of an aëriform body to spread and diffuse itself, is spoken of as an "extreme elasticity," and it is accompanied by an extreme degree of *compressibility*. Both these extremes are characteristic of aëriform bodies exclusively. Nevertheless, like solids and liquids, they possess *weight, inertia, momentum*, and *impenetrability*—not, of course, that a *mass* of air is impenetrable, and we may, with a finger, penetrate even into the mass of a soft solid! But the real essential substance of aëriform, as of all other, matter, is deemed to be impenetrable in the sense that it must always remain of some dimension and cannot be made actually nothing of.

Aristotle was aware that air was a material substance and, like other bodies possessing weight, tended to descend towards the earth. In fact its weight is very considerable, and greatly modifies the circumstances and actions of liquids, so that some additional facts about the latter will have to be noted before concluding this chapter. Such is the case because, in treating of liquid bodies, we made abstraction of the action of aëriform ones as being things which had not yet been brought before the reader's cognisance.

The weight of the atmosphere—*i.e.*, of the aërial mass round the surface of the earth, at the sea-level—is between fourteen and fifteen pounds upon every square inch. We say "at the sea-level," because it is obvious that the more we ascend above this level the less will be the volume of air which presses downwards. But the decrease in pressure is very far from being uniform, because the lower strata have to bear all the weight of the strata of air above them; and the aëriform bodies being exceedingly compressible, the lower strata are much the denser, and density rapidly diminishes upwards.

The body of an ordinary man has to sustain a pressure of about 33,600 pounds or 15 tons; but we are no more inconvenienced thereby than is a fish by the pressure of the ocean, and for the same reason.* The property which both water and aëriform substances possess, of pressing equally in all directions, serves no less to sustain than to oppress. Thus the most delicate glass or other vessel is enabled, without the slightest injury, to sustain atmospheric pressure. If, however, by any contrivance the air within such a vessel be removed, then the vessel will be immediately crushed.

There are machines to remove air from a vessel, as well as to force more into it—such as an exhausting syringe (for the former process) and a *condensing syringe* (for the latter purpose). By means of the condensing syringe, air may be added to what was already contained, e.g., in a copper flask of 100 cubic inches capacity. Thereupon the flask will be found to have increased in weight. If, on the contrary, the air is removed from such a flask by means of an exhausting syringe, it will weigh 31 grains less, so that the weight of an ordinary 100 cubic inches of air must be 31 grains.

Another instrument for this purpose is called an air-pump, and by it air can be removed from the interior of a large and strong glass-vessel called a "receiver." If before it is emptied (so far as it can be emptied) of air, a delicate glass-vessel, perfectly closed and containing air, be put within it, then, as the air is removed from

---

* See *ante*, p. 70.

the receiver and so the external pressure, antagonistic to that of the air contained within the delicate vessel, ceases, the vessel will be blown to pieces by the unopposed elastic force of the air which it contained.

Thus it is plain that any small portion of air so cut off from communication with the atmosphere, still exercises pressure (a pressure that decreases as the volume becomes greater), which cannot be due to the weight of so small a portion of air, but must arise from its expansive force alone.

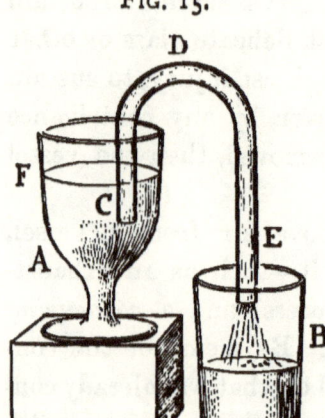

FIG. 15.

As has been said, the laws of pneumatics modify in various degrees the actions of liquids. Of two vessels, A and B (Fig. 15), let A be filled with water up to the level F, and B be empty. Then let a bent tube CDE (the limb DE being longer than the limb DC) be filled with water and temporarily closed (*e.g.*, with the finger) at either end. Next let the bent tube be so placed that the end C be immersed in the liquid at A, while the other end, E, is over the empty vessel B. Finally, let both ends of the tube be simultaneously unclosed. Then as gravity brings down the water in the limb DE it will be replaced by a rise of water from A up the limb CD to replace it—a rise due to the pressure of the atmosphere on the surface of the water in A. By the continuation of this process the water in A will gradually become transferred into the vessel B. Such an instrument is called a *siphon*. The same action will result if the tube be placed in a similar position empty of water, provided only it is also emptied of air, either by an exhausting syringe or by the action of the mouth in sucking. When it is thus emptied, the pressure of the atmosphere on the water in A immediately causes it to rise; because there is no longer any pressure within the tube to counteract that of the atmosphere over A, the air in the tube having been removed. *Suction*, or the action of *sucking*, essentially consists in the withdrawal of air, followed by changes induced in consequence through atmospheric pressure. The immersed limb of the siphon must be shorter than the other, in order that the resultant pressure of the liquid in the tube, and of the atmospheric pressure may act in the direction CDE. Were both the limbs of the tube of the same length, the atmospheric pressure at either end being equal, the water would then simply fall back into the vessels A and B.

In a boy's squirt the principle of suction is brought most simply into operation, as also in the common household pump. In the common pump there is a valve at the bottom of the space in which the piston works, and this opens and allows the water to ascend (through the tube which dips down into it) up to that space and so fill the partial vacuum produced by the ascent of the piston. Another valve in the piston opens to allow water to ascend towards the spout, together with any air which may be left to escape, while the lower valve simultaneously closes, and so prevents the re-descent of the water previously raised.

It was long supposed that this ascent of water was due to the production, by pumping, of an absolute vacuum, which being a thing Nature abhorred, water rose spontaneously to fill it. Even Galileo thought it was due to an attraction exercised on water by the piston. A deep well at Florence having failed to draw water, the attention, first, of the more illustrious Italian just named, and afterwards of his disciple Toricelli, led to the discovery that the pressure of the atmosphere will not counterbalance the weight of a column of water more than between 33 and 34 feet high. Experimenting with mercury, the last-named observer found that after filling a tube, three feet long and closed at one end, with mercury, and immersing its open end in a vessel containing that fluid, the mercury in the tube sank till it stood about 30 inches higher than the surface of the mercury in the vessel. The height of the mercury sustained by atmospheric pressure was found to be so much less than that of water, on account of the much greater density of the mercury. The space left above the mercury which had so descended was supposed to be absolutely empty, and is known (on account of the name of its observer) as the "Toricellian vacuum." This space is, however, really filled with the vapour of the liquid. Barometers are tubular instruments which measure the weight of the atmosphere by showing the height to which the pressure of the air will raise a column of mercury (or other fluid) contained within them. As it is evident that the higher we ascend above the earth's surface the less the weight of the atmosphere will be, it is no wonder that barometers serve to indicate height, when once their condition at the sea-level has been accurately ascertained. Barometers only serve to indicate approaching changes of weather in so far as such changes are connected with a denser or lighter atmosphere. That aëriform bodies may attain a great momentum and exercise a vast amount of pressure is shown by the effects of cyclones and hurricanes. Cyclones are rotatory movements of air, which may be readily occasioned under certain atmospheric conditions, on principles similar to those by which, as we have seen, rotatory movements of water may be produced.*

As in liquids, so in aëriform bodies, movements may take place in the form of waves. This is shown by the oscillating movements of the atmosphere near the surface of the earth on a hot, sunny day in summer. Such oscillations are made manifest by the apparent twinkling movement of the objects seen through the oscillating waves of air.

But these subjects, heat and the light which makes things visible to us, are subjects which pertain not to mechanics, but to those sciences, the elements of which will be briefly introduced to the reader's notice in the next chapter.

Like liquids, aëriform bodies differ much in density. This it is which makes a balloon rise in the air as bubbles of oil will rise in water. The reason why a balloon rises is that it contains an aëriform fluid so much lighter than air that its whole weight is less than that of the bulk of air it displaces, and thus the relatively heavier air descends, and so presses it upwards.

---

* See *ante*, p. 74.

CHAPTER IV

PHYSICAL FORCES

The most universal properties—number, figure and motion—possessed by all those bodies which our senses can take cognisance of—whether such bodies are solid, liquid, or aëriform—have now been treated of in an elementary manner.

We may next proceed to consider those forces which are commonly said to affect bodies—forces which bodies, at any rate, make manifest to us, more or less frequently. Every one knows that water behaves differently at different temperatures, and that the air is greatly affected by heat, as also that the same is true of solid substances, though in very diverse degrees. We all know that some bodies are, or can be rendered, luminous, as well as that sounds are transmitted to us through the air. Some readers may have seen sparks emitted from the hairs of a cat when rubbed, while others have doubtless, while children, amused themselves with magnets, and no one can be ignorant that the cleanest iron will get rusty when long exposed to the air.

It is also a matter known to everybody, that one and the same material substance will be now hotter, now colder; now brightly luminous, at another time dull (as *e.g.*, a coal); occasionally sonorous and other times silent (as a piano); now tranquil and motionless, yet afterwards conspicuously turbulent—as two effervescent powders will lie perfectly motionless when mixed together in a dry state, but become violently turbulent, when together thrown into a glass of water. Every one knows that such changes are but transitory, as well as that a telegraph wire, however much used, is not perpetually active; and some readers (who have attended scientific lectures) may probably have learned that a body may be magnetic or not magnetic, according to circumstances.

Thus a conception has very naturally arisen that these manifestations of activity in material things—activities which we will speak of as physical forces—are entities or influences, which come and go—which pervade bodies for longer or shorter periods, and then leave them—as (to use a rough-and-ready illustration) a sponge may be soaked full of water and squeezed dry again, any number of times.

The present chapter will be devoted to an exposition of some elementary facts concerning these pervading influences, the energies—or "forces" to continue the use of a popular term—known as *heat, light, sound, chemical change, electricity,* and *magnetism.*

It is evident that bodies which every now and then exhibit any one of these forces, continue to possess, at times when they do not exhibit it, the power of again manifesting it when the necessary conditions return. The energy under such circumstances is said to be in a "*potential* condition," as distinguished from an active, or as it is technically termed, "*kinetic,*" state. Such potential energy is a capacity for a certain activity—*e.g.*, doing a certain amount of work—which capacity is actively expended in overcoming some definite resistance—*e.g.*, overcoming it through a definite distance—and is therefore capable of measurement. The term "force" denotes that which is the cause of all and every known kind of motion. As there are the six kinds of active energy above mentioned, so we may speak of their unknown cause as so many "physical forces"—in a quite elementary work, such as the present one, questions as to the absolute nature and distinctness of such forces cannot be touched upon, though conceptions which serve science as working hypotheses, will be referred to.

We have already made acquaintance with the force spoken of as gravity, but here we shall not further study the energy due to that force. For it is a universal, constant condition of all material bodies, constant, moreover, not only in its existence and action, but also in the precise amount of its action, which is ever in exact proportion with the mass of which any body (its distance from other bodies being unchanged) consists—as has been pointed out in the last chapter.*

We have noted the various modes of motion and tendencies to motion in solids, liquids, and aëriform bodies, but we all know that one substance at least can exist in all three conditions. We know that water can be both frozen and changed into a vapour; while steam and every other vapour is an aëriform body. Steam is an invisible vapour. What is popularly called "steam" is a cloud of minute particles of water—formed by the resumption of its liquid condition. If we look at the place where steam is issuing from a rapidly boiling kettle or engine, we shall find that nothing is visible close to the mouth of the spout or chimney, the cloud of what is popularly known as "steam" only begins to appear at a short distance from it.

HEAT.—It is notorious, as before said, that cold

---

* See *ante*, p. 66.

occasions the assumption by water of its solid form, as ice, and that heat will convert it into steam. By the continued application of heat, a vessel of water may be emptied—the whole mass being boiled away into aqueous vapour, while when that vapour passes into a cooler space it becomes condensed into a cloud of particles of water—as above stated.

A multitude of other substances are known to be capable of similar changes as, for example, mercury. It is notorious that lead can be easily made liquid by heat, and iron also, though not as readily. There is little doubt but that all substances can exist in these three states. Various gases have been made both liquid and solid, and Professor Dewar has recently liquefied the gas oxygen and even the air we breathe, while many components or rocks have been rendered both liquid and aëriform.

Heat then is evidently a very powerful agent in effecting the change of state from solid to liquid, and also that from liquid to vapour. Pressure has a contrary tendency, but it requires an enormous amount of pressure to counteract the effect of a very little heat. In mechanics, we have regarded bodies as incompressible, but in fact they are all compressible, though in very different degrees. Aëriform bodies are easily compressed, and when released from pressure, spontaneously expand; but extraordinary force is required to compress water, and greater force still to compress solids, which are almost incompressible.

Liquids may spontaneously pass into the aëriform condition. Such is the case with water, a thin layer of which will (with different degrees of rapidity according to circumstances) "dry up" by a spontaneous process, called *evaporation*, by which it passes into the state of vapour as in boiling or "ebullition." Pressure produces a definite effect on this process. Thus the weight of the atmosphere causes water at the sea-level to need a greater supply of heat to boil than is required on a very elevated mountain, where its pressure is necessarily much less. By lowering temperature, the vapour will again assume the liquid condition — as before said.

But heat does not only act as a transformer of bodies from one state to another, it also exercises one very notable effect on bodies which remain in an unchanged condition, whether that be solid, liquid or aëriform. It makes them expand, as we see in the thermometer, wherein heat causes the liquid it contains to rise in a vertical tube, owing to the expansion it produces in that liquid. Heat applied to a gas enclosed in a vessel, cannot, of course, expand it (beyond what may be allowed by the expansion of the vessel itself from heat) on account of its boundary, but by its very gaseous nature, it is always expanded, and presses upon every part of the vessel enclosing it, however low the temperature—supposing it is not low enough to turn the gas into a liquid. But it will tend to make the gas press more energetically against the vessel containing it, and may cause that vessel to burst.

The whole world is continually and everywhere under the influence of heat, and when heat is passing from one body to another, the former is said to be of a higher "temperature" than the latter, the temperature of which is raised by the heat it receives from the former. Thus the terms hot and cold imply a relation existing between two bodies as regards their temperature. There is absolutely no such thing as absolute coldness; cold is but a relative term.

Though bodies expand,* or tend to expand, through the agency of heat, it is very evident that different bodies and substances do not expand at the same rate. Thus it is plain that both the mercury in a thermometer and the glass vessel which holds the mercury do not both expand equally, or the mercury would not rise in it as it does.

Liquids expand more than solids, but different liquids as well as different solids expand at different rates.

To estimate differences of temperature it is necessary to adopt a definite external standard. This is necessary because not only are our feelings insufficiently persistent to enable us to use them as a test, but (for reasons to be explained shortly) they may positively mislead us as to the relative temperatures of bodies we successively touch.

The instrument made use of is, of course, the thermometer, which is marked in a manner agreed upon, so that it may serve as a standard of comparison. It has been ascertained that the temperature at which any liquid becomes solid is always the same, as also the temperature at which any fluid boils—the conditions under which the ebullition takes place being similar. It is this fact which enables a thermometer to be graduated —the freezing and boiling points of water being thus constant. In England the arrangement adopted is that called the scale of *Fahrenheit*, according to which the space between the position of the mercury in the tube when the thermometer is plunged into melting ice and that at which it stands when in boiling water, is divided into 180 spaces or "degrees." The space below the freezing point of water is divided into 32 similar spaces; and thus, according to this system, the freezing point of

---

* As to water, see *post*, p. 91.

water is 32° and its boiling point 32 + 180, or 212°. The lowest point of this scale is called "zero," and degrees below this, are spoken of as so many degrees below zero. The degrees between zero and 32° are also spoken of as so many "degrees of frost." Thus 20° marks twelve degrees of frost. On the Continent, *Centigrade* thermometers are used, according to which the space between the points of freezing and boiling water is divided into 100 degrees, and the freezing point is the zero of that system.

Now if various bodies of different kinds, all have their temperature simultaneously changed to the same extent (all made 10° hotter or colder) they will expand differently; that is, the ratio of the change of bulk will be different for every different substance.

Different substances have, indeed, very different capacities for heat, and the same amount of heat, communicated to two different bodies whose masses or weights are equal, will not cause the same rise of temperature in each. Thus a pound of water at 40° and a pound of mercury at 160°, if mingled together, will not produce a mass of a temperature of 100°, but only one of 45°, the temperature of the water thus having only risen 5°, while that of the mercury has fallen 115°.

If we take equal masses of different bodies, A, B, C, D, &c., then the numbers which are proportional to the various amounts of heat required to make them all of the same temperature are called the *specific heats*, or the *capacities* for heat, of A, B, C, D, &c., respectively.

But, as before said, heat not only expands bodies, but, if continued, will change solids into liquids and liquids into gases. Sometimes, but by no means always, solids will, in melting, pass through an intermediate, or *jelly-like*, condition before becoming liquid. A jelly is a substance which, for scientific purposes, may be conceived of as being incompressible—as we suppose solids to be—while its particles are incapable of sliding freely over each other, there being a partial resistance thereto.

With these changes of state, changes of volume do not always correspond. As a rule, a solid increases in volume in liquefying, and again increases in assuming the aëriform condition, while it shrinks as the process is reversed. With water, however, it is not so. In becoming solid (ice), it increases in bulk, and becomes lighter than in the liquid condition.

But there are certain notable facts with regard to heat in relation to such changes of condition. Heat curiously disappears during changes from solidity to liquidity, and from liquidity to gaseousness, but reappears in changes from gaseousness to liquidity, and from liquidity to solidity.

Ice will absorb a larger amount of heat without indicating any rise of temperature, until the whole of the ice is melted. The heat so absorbed is commonly called *heat of liquefaction* or *latent heat*, as distinguished from heat which makes itself manifest. It is also called *potential heat*, because it can reappear. It is thus again *evolved* and "reappears" when a substance passes from a liquid to a solid condition.

If water be boiled, and so raised to a temperature of 212°, it will not show any higher temperature while exposed to ordinary atmospheric pressure, even though an amount of heat be applied to it enough to raise it 970°. In this way much heat again disappears and becomes potential, reappearing and becoming actual and energetic once more, when vapour condenses into the fluid condition.

Mercury freezes at a temperature 38° below zero, or

−38°. Ice (as before said) melts at 32°, but tin needs a temperature of 442° to melt, and antimony requires 812°.

We have just seen that heat can disappear and become potential, as also that it can again reappear or be evolved. It can also be—(1) conducted, (2) conveyed, (3) radiated, (4) absorbed, (5) reflected, and (6) refracted. The real nature of heat in itself is absolutely unknown, that is to say it is only known through certain effects—itself remaining permanently inaccessible to our senses.

Various suppositions, or hypotheses, have been suggested as to its nature, the value of which depends on the extent to which any of them can enable us to harmonise, foresee, and predict different actions of various bodies, which are directly perceptible by us.

It was at one time supposed to be a peculiar kind of substance termed "*caloric.*" But such substance, it was evident, could have no weight, since nothing is made heavier by being heated, or having, as it was supposed, "caloric" added to it. It could also exist without manifesting itself in any way—in the condition just above spoken of* as "latent" or "potential" heat.

But another hypothesis found favour subsequently; an hypothesis which, it will be seen, is still more useful in reference to the next physical energy we shall consider —namely, light. We have said† that in mechanics it is convenient to suppose each body to consist of a great quantity of minute particles, and the different supposed motions of these supposed particles are taken to explain the diverse conditions we describe as liquid ‡ and aëriform.§ Now we may suppose these particles to be, as it

---

* See *ante*, p. 91.   † See *ante*, p. 43.
‡ See *ante*, p. 68.   § See *ante*, p. 78.

were, immersed in another sort of fluid substance of an indefinitely more refined nature, called *ether*—a substance tending to separate the particles (and therefore to oppose gravity), and also possessed of and able to transmit various orders of vibratory motions. Motion so unimaginably minute taking place amongst the particles, or "molecules," of bodies is termed *molecular motion*. It is thus distinguished from the motion of bodies we can perceive, which is called *molar* motion, or the motion of perceptible masses of matter.

Heat is now treated as if it consisted of such molecular oscillations, which are conceived of as varying in extent and velocity, but as continuing perpetually and tending to become everywhere uniform—by intercommunication between all particles of all bodies without limit. Such intercommunication is supposed to take place through the hypothetical refined substance, called ether, which is conceived of as being something essentially different from the solids, liquids, and aëriform bodies, of which we have hitherto spoken, and also conceived of, as being universally diffused. For some scientific purposes, it is most convenient to suppose this ether to be an ideally perfect fluid; while for others it is treated as an ideal jelly-like substance.* Men of science are of course quite free to treat it in any fashion which may help on investigation, but we must not regard such speculative hypotheses as representing real, ascertained truths. Moreover, as before said, an elementary work like this is not the place to treat of such a problem as the question what "heat" in itself may be. We must be content with serviceable hypotheses. One such is that it consists of molecular vibrations varying in intensity and capable of propagation through space in

---

* See *ante*, p. 90.

all directions. Some facts with regard to heat accord best with that conception which regards it as a peculiar kind of fluid. Other facts fit in best with the hypothesis that it consists of molecular motions; while yet other facts seem to require a union of both these hypotheses. Such matters, however, can be little more than glanced at here, though it is necessary to bring some representation of molecular motion and of ether, before the mind of any one who desires to become acquainted with the elements of science. This is the more indispensable because there are definite quantitative relations between molar motion and the molecular motion of heat. The heat produced in iron by the strokes of the smith's hammer, and the occasionally setting on fire of wheels—produced by their revolution (when friction much impedes motion)—roughly show this; but careful investigations have now revealed precise numerical equivalents between heat and the motions of bodies.

Leaving all hypotheses on one side, we will now consider the six various phenomena of heat we enumerated before we began to speak of heat in itself.

(1) *Conduction.*—When two bodies of different temperatures are placed in contact, the hotter body becomes cooler, and the cooler body becomes hotter, till both are of the same temperature. When they are bodies equal in mass and of the same substance, the increase of temperature in the one will be equal to the decrease of temperature in the other. Heat is thus conducted from one body to another. This transference of heat takes time, but the time varies greatly according to the nature of the substance; some substances being much better conductors of heat than others are. Thus even a short piece of charcoal burning at one end can be held in the fingers at its other end without inconvenience;

but a short piece of iron made red hot at one end cannot be so held: iron being a much better conductor of heat than charcoal. Wood, woollen substances, and fur, are notoriously bad conductors of heat; while metals are notoriously good conductors of it. Gold conducts much more than twice as well as iron. If the conducting power of gold be taken as 1000, iron is as 381, while that of marble is but 23, and clay only 11.

We have before spoken* of the insufficiency of our mere sensations as measures of temperature. If the hand be plunged in water, really as warm as the air, the water will feel colder. If also the hand be placed first on fur, then on a wooden table, and then on a marble one, the last will feel the coldest. This is because the marble is a better conductor than the other substances, and so conducts heat more quickly out of the warm hand. For the same reason metal will feel still colder. It cannot safely be touched in the Arctic regions with the naked hand, because it conducts heat so rapidly from it as in effect to burn it. If wood and metal be both made equally hotter than the hand, the metal will feel much the hotter of the two, because it will conduct heat into the hand much more quickly.

(2) *Convection.*— In a solid body, heat passes through it without occasioning any change of position between its constituent parts so long as the particles cohere, but in liquids it is quite otherwise. On account of their extreme mobility some of the particles which compose them are displaced by every change of temperature—the warmer particles ascending and so conveying the heat towards other parts of the liquid mass, while by so doing they necessarily displace

---

* See *ante*, p. 89.

colder particles which, being heavier, take the places of those which have been expanded by heat, and have so become specifically lighter.

In aëriform bodies, convection also takes place; any hot body causing upward currents at once. This is the cause of that apparent twinkling movement of objects before spoken of * as often to be observed, on a brilliant hot day, immediately above the surface of the ground. It is produced by waves of air of different temperatures, and therefore of different densities, which refract † the rays of light, and so modify the appearance of objects seen through such waves of air.

The greater the power of convection a body possesses, the less its power of conduction. Thus water in a closed tube with a piece of ice at the bottom of it may be made to boil at the surface while the ice will remain unmelted. Aëriform bodies are even worse conductors than fluids are.

(3) *Radiation.*—It is a most familiar fact that we can very quickly obtain much warmth by standing in front of a bright open fire or a mass of red-hot embers. This heat is certainly not obtained by conduction, seeing how bad a conductor of heat air is. It is also not due to convection, for the effect is too instantaneous and the hot air (displaced by the rush of cold air towards the bright and glowing mass) would not be conveyed horizontally outwards, but upwards to the space whence the cold air had descended.

The fact is that the incandescent or glowing mass gives forth what are called *rays* of heat. These "rays" may be supposed to be—(*a*) an influence, or (*b*) a mode of motion, or (*c*) a substance emitted by the fire and

---

* See *ante*, p. 83.   † See *post*, pp. 104 and 105.

extending towards us when we seek warmth by standing near it. But these rays are not only given forth horizontally; they are radiated in all directions—as may be proved by suspending a red-hot sphere, when the rise in temperature produced by it will be found to be equal on all sides of it. The rays thus given off proceed in straight lines.

The power or intensity of heat thus radiated is like the force gravity,* in that it varies inversely as the square of the distance. It is four times less at two feet distance than at one foot, nine times less at three feet,

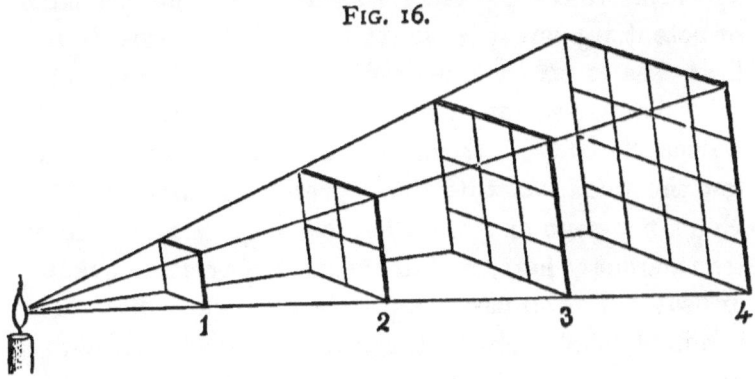

Fig. 16.

sixteen times less at four feet, and so on. This is because the heat-rays radiating from any point spread out at a distance of two feet to four times the extent of space they do at one foot, to nine times at three feet, sixteen times at four feet, and so on.

All bodies are constantly radiating away their heat, and as constantly receiving it from other bodies, but they do this at unequal rates—*i.e.*, their radiating force is unequal. This is the case even when different bodies are at the same temperature. Thus mercury has only

---

* See *ante*, p. 66.

one-fifth of the radiating power of lamp-black, which is almost equalled by writing paper; while polished gold, silver, or copper has but little more than half that of mercury. But the rate of radiation has more to do with the state of the surface of a body than with the nature of the material whereof it is composed. Bright surfaces radiate least, but their power will be almost doubled if their surface be covered with lamp-black.

Practically the rate of diminution which distance occasions is more or less increased by heat being absorbed.

(4) *Absorption.* — We have already spoken of heat appearing to become absorbed when it is rendered latent or potential; but it is said to be truly "absorbed" when heat passes from the radiating to the conducting condition. As radiant heat traverses a body, some of it warms the body it traverses, and becomes conductable, but the amount of this differs greatly in different bodies. None, not even air, are what we may term absolutely transparent to heat, or *diathermous,* but a certain quantity of heat will often pass completely through bodies. This is notably the case with the atmosphere, which allows so great a quantity of the sun's rays to traverse it without warming it, that almost the entire quantity comes to the surface of the globe, which, being thus warmed, gives to the air, by convection, that heat which it failed to receive from the heat radiated through it. Bodies absorb heat at precisely the same rate as they radiate it. There appears to be some ground for supposing that rays of heat differ amongst each other by some other quality besides intensity; since some rays are more absorbable than others, and so become filtered out first, while heat is passing through some more or less diathermous medium. The absorbability of heat differs then with respect to different media, as we have just seen.

(5) *Reflexion.*—When considering radiant heat it is convenient, since it always proceeds in straight lines, to think of it as divided into an indefinite quantity of straight lines or rays of heat—as we before found it convenient * to represent the action of gravity by a number of parallel lines.

As in dynamics we found that when a ball impinges on a surface its angle of incidence is more or less (according to its elasticity) equalled by its angle of rebound, or of reflexion, so rays of heat will rebound from a surface according to the same law—namely, that both the lines of impact and rebound must lie within an imaginary plane perpendicular to the reflecting surface at the point of contact.

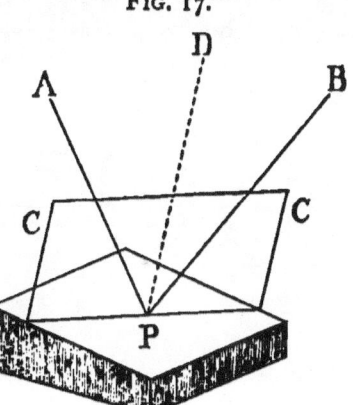

FIG. 17.

Thus if a ray A (Fig. 17) falls at the point P, on a reflecting surface, it will be reflected to B, and a line DP perpendicular to the surface D will make the angle DPB equal to the angle DPA and the perpendicular, and both the incident and reflected rays, will all lie in the plane CC.

When two concave reflecting surfaces—mirrors—with a certain definite curvature, are placed opposite each other, a very curious effect may be produced.

At a certain distance from each mirror is a spot called its focus, all the rays radiating from which to the surface of the adjacent mirror, will be reflected in parallel lines. These rays impinging on the surface of

---

* See *ante*, p. 43.

the mirror opposite, will be so reflected as to meet again exactly at the focus of the second mirror—the spot to which all the rays coming to the mirror in straight lines are reflected and converge. Now, if a red hot iron ball be placed at A (Fig. 18), the focus of one, the rays of heat will then radiate from the adjacent mirror's surface, thence they will be first reflected in straight lines to the surface of the other mirror, whence they will be again reflected in convergent lines to its focus B—when if either a piece of phosphorus or a thermometer has been previously placed there, the thermometer will rise or the phosphorus take fire. The interposition of a screen between the mirrors will prevent these effects,

FIG. 18.

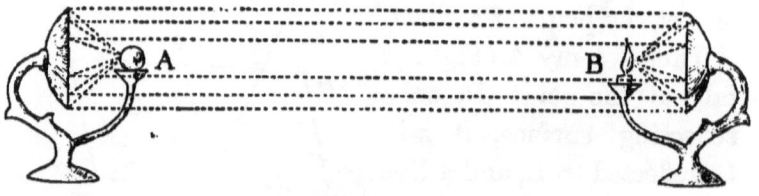

while if there be no screen, the red hot ball may be placed even nearer the thermometer than the focus of the first mirror and yet produce much less effect than when in that focus.

(6) *Refraction.*—When radiant heat passes from one medium into another—as from air into glass—then the directions of its rays become thereby somewhat changed, or, as it is called, *refracted*. If the second medium be diathermous (transparent to heat) then, when the rays, having traversed it, pass out of it again—*e.g.*, from glass once more into air—there is a second refraction of the rays which completely undoes the effect of the first as to direction—if (as in the case supposed) the third

medium be similar to the first, and the opposite surfaces of the intermediate medium be parallel.

If, however, both its surfaces be curved in a convex manner, parallel rays falling through one convex surface will converge, and will converge again when passing

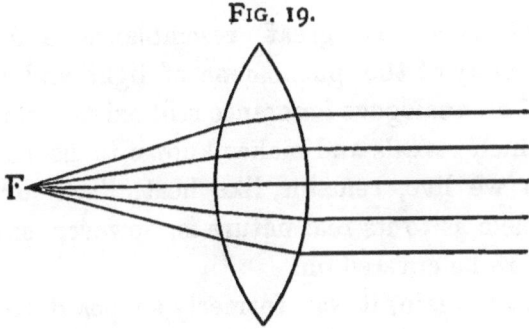

FIG. 19.

out of the opposite convexity till they all meet at a focus F (Fig. 19).

Thus a doubly convex lens will bring the hot rays of the sun to a focus where they may set matter in combustion—the lens acting as a " burning glass."

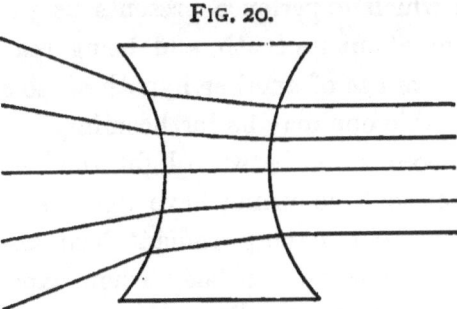

FIG. 20.

Doubly concave surfaces (Fig. 20) will, in an analogous manner, increasingly scatter and disperse the heat rays.

The refrangibility of heat differs according to the medium it traverses, and this is another indication that heat rays differ by some other quality than mere intensity.

It has been ascertained that some rays of heat which have been reflected, or refracted, are not equally energetic in all directions to which they may (by turning the reflecting or refracting body) be directed; but this phenomenon will be best considered when we treat of light.

Light.—There is great resemblance and accord between many of the phenomena of light and those of heat, and an analogous ignorance still exists about both. Light, which reveals and makes known to us the world in which we live, remains, like heat, itself unknown. The problem as to its real nature is, however, one which cannot here be entered on.

Like heat, again, it was formerly supposed to consist of minute material particles emitted (by luminous bodies) in all directions, but afterwards it was regarded as the effect of minute vibrations, or waves, of an elastic, universally diffused ether. All that we need do here is to welcome, as a working hypothesis, that representation which best helps us to understand and anticipate the phenomena which experience presents us with, keeping an open mind about its truth, and being ready to lay it down and make use of another hypothesis so soon as any more serviceable one may be forthcoming.

The close connection between light and heat is obvious from the fact that we cannot have light (*e.g.*, from fire or candle) without having radiant heat also, and the warmth to be felt by our body when exposed to the brilliant light of the sun tells us the same thing.

It is abundantly evident that whatever light may be in itself, like heat, it diffuses itself in straight lines in all directions from every visible object, and, of course, it also comes in straight lines to the eye from every visible object.

The light emitted from any point of any object is called

a "*pencil*," and is said to consist of "*rays*" of light, and these can cross each other in the same point of space without either hindrance to their action or even diminution of their intensity. This may be shown on a vertical white surface opposite a minute aperture admitting the rays of the sun into a room otherwise quite dark. Then any external object—*e.g.*, a tree—will be represented upside down on the white surface. The rays of light emanate from every point of that tree in all directions, but none but those from its upper part can reach the bottom of the surface opposite the small hole, while none but those from the bottom of the tree can ascend towards the top of that surface, and the same consideration applies to all the rays emitted from each intermediate point of the tree's surface. Therefore any object, thus viewed, must appear inverted.

Light radiates as heat does, and its intensity also varies inversely as the square of the distance, a rate practically diminished by the absorption it may undergo in traversing bodies—as, *e.g.*, the air. Bodies which light can traverse are called *transparent*. Those which entirely absorb it are *opaque*. Bodies which are opaque may sometimes be rendered transparent very easily. Such is the case, *e.g.*, with a kind of agate known as *Hydrophane*. In its ordinary condition it is only half-transparent, but can be made perfectly so by immersion in water. Similarly, paper by being oiled, can, as the reader knows, be made much more transparent than before.

Light travels with amazing velocity—at not less a rate than 186,330 miles a second, yet this does not prevent what is called the "*aberration*" of light, which will be explained * when we come to the

---

* See *post*, p. 176.

consideration of bodies very distant from the earth's surface. Light can travel with perfect freedom through any glass vessel, the interior of which is as perfect a vacuum* as we can possibly make, therefore if ether is necessary for the existence of light, ether must fill any such glass vessel, and therefore its interior can be no real vacuum. Like heat, light is eminently capable of reflexion,† wherein it may be said to follow in a general way the laws which determine the reflexion of heat. The quantity of light reflected varies greatly, and rarely amounts to one-half—so much being generally absorbed. Any object seen appears the darker, the greater the quantity of light thus absorbed. All surfaces equally reflect light, but those which are "dull" are minutely rough—*i.e.*, have minute surfaces turned in many different directions and these reflect light equally in all possible directions. When a body appears "glittering," it is because it bears a quantity of small surfaces, or facets, which are nearly smooth, and therefore each such facet reflects light similarly, and for the same reason a polished surface—as that of a mirror—appears to reflect light most perfectly. It reflects so perfectly, because its smoothness causes it to reflect with order and regularity—according to the law of equal angles of incidence and reflexion—the rays which reach it with order and regularity from neighbouring objects. There are certain phenomena of reflexion connected with colour which will be best considered together with the next property of light, namely *refraction*.‡

Rays of light, like those of heat, are refracted as they pass from one medium into another. Light falling on a

---

* See also page 60. † See *ante*, p. 99.
‡ See *post*, p. 108.

transparent medium is in part reflected, while the main part which traverses it becomes refracted on the way. It is this refraction of the rays of light which pass to the eye from all the parts of a stick partly immersed in water, which causes the stick to appear bent at the point where it enters the liquid, and also causes any solid object placed on the bottom of an empty vessel to appear to change its place when the vessel has water poured into it.

This is due to the fact that such rays of light do not

FIG. 21.

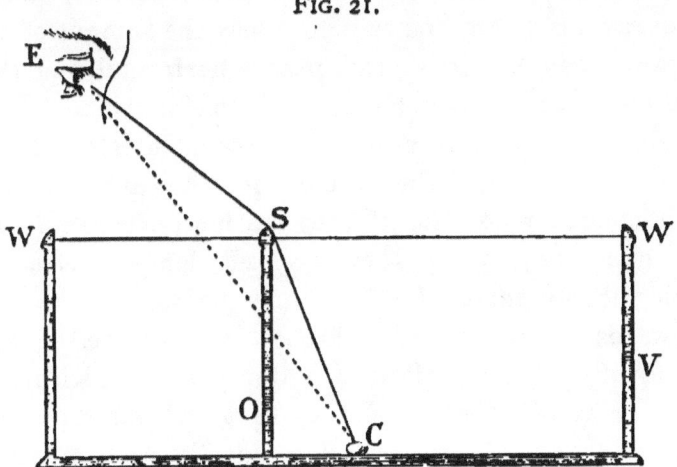

proceed in a straight line to the eye from the object looked at, but are deviated, bent or refracted, from the point where they impinge upon the surface of the second medium, as in passing from the water to the air. Thus it is that refraction may enable us to "look round a corner," because a ray of light proceeding from out the water is bent, by refraction, so as to be more nearly horizontal to the water's surface.

Let E (Fig. 21) be the position of the observer's eye, C that of a coin lying on the bottom of a vessel V, and the dotted line CE be the direction of a ray passing straight between E and C. Then it is evident that if an opaque vertical partition or septum, such as O, be interposed in the course of that straight line, the coin will be invisible. If however the vessel be filled with water a ray passing from C to S, just above the summit of the septum, will be refracted to E for the following reasons. A ray passing from air into water always becomes more nearly perpendicular to its surface and one passing from water into air, always, as before said, becomes more nearly horizontal to its surface, and so the ray CS (from the coin to above the summit of the septum) will be made more nearly horizontal and thus pass to E and cause the coin to become visible.

Rays which fall, or ascend, perpendicularly, are not refracted at all, but the more oblique they are, the more refracted they become, till they reach an extreme degree of obliquity; when they are no longer refracted, but entirely reflected. This may be seen by looking upwards or downwards through a glass vessel, very obliquely, at the surface of water contained within it. The water will then appear to have lost all its transparency, and will reflect as an ideally perfect mirror would do. For the laws of refraction the reader must have recourse, as for all that is not quite elementary respecting physical forces, to professed treatises on physics.

Light passing through differently shaped transparent media exemplifies supremely well the laws previously stated with respect to the passage of heat through bodies with differently shaped surfaces. When light passes through glass which is flat (*i.e.*, the opposite surfaces of which are parallel) it is of course doubly refracted—the refraction undergone on entering the glass being reversed when it emerges from it so that it regains its original direction, although there is a slight change of position. If the opposite sides of the glass are not parallel then that piece of glass is what is called a prism. By it the direction of the rays is permanently changed, and the more so, the more inclined to each other the two surfaces of the glass may be. It has also certain other effects, respecting colour, which will be referred to a little further on.

The light of day having so very distant a source, may be practically regarded as consisting of parallel rays, just as the force of gravity on the earth's surface may, as we before saw,* be conveniently treated as if acting in parallel lines, although really it acts in lines radiating from the earth's centre. Rays of light, then, which fall upon and pass through bi-convex or bi-concave bodies, follow laws similar to those which we have already seen apply to rays of heat.† Thus it is that by a judicious combination of glasses, those instruments, so valuable for Science, the *microscope* and the *telescope*, are constructed. Thus also an image of external objects can be made to fall upon the surface of a table in a camera obscura. Bi-convex glasses, or lenses, bring the rays to a focus at different distances according to the curvature of their surfaces, and the distance in each case is called the *focal length* of such lens.

When treating of heat, but very little could be said about refraction. But now, after what has been said about light, it will be easy to understand how and why rays should thus be made to converge or diverge according to the shape of the surface, whereon they impinge. For however many parallel rays fall upon a curved surface, they must all have different inclinations to it,

---

* See *ante*, p. 43. † See *ante*, p. 101.

and must therefore undergo correspondingly different amounts of refraction. Therefore it is impossible they should be parallel when they issue forth from a body after having entered it through a curved surface. Every convex surface causes convergence of the rays, and every concave surface scatters.

When speaking about heat, we said* that there appeared to be some reason for supposing that its rays differ qualitatively as well as in mere differences of intensity. What was thus suggested as to heat, is certain and evident as regards light.

No reader can have failed to see, now and again, manifestations of colour in the neighbourhood of some piece of glass, or of crystal, and he may have remarked a resemblance between the colours thus appearing and the hues of the rainbow. In the rainbow, the lowest of the series of tints it exhibits is violet, and to that succeed indigo, blue, green, yellow, orange, and red. There are, however, so many intermediate shades, that it is impossible to see where one ends and the next begins. It is generally believed that light, which is apparently colourless, somehow consists of different coloured rays. This belief certainly appears to be confirmed by a very simple experiment. If a circular piece of cardboard be painted with the colours of the rainbow, each patch of colour narrowing to a point at the centre of the card; then if a rod be passed through that centre, and the card be turned rapidly round it, the separate colours will disappear and the card will assume a grey, or nearly white, appearance.

Colour has been before referred to,† but its consideration was postponed until the reflexion of light

---

* See *ante*, pp. 98 and 101. † See *ante*, p. 104.

came to be spoken of, because the hypothesis now popularly employed to explain the different colours which objects present to us and seem to possess, explains them by reflexion—colours being represented as due to different conditions of the reflexion of light. Light being normally colourless, or white, a white object is one which is supposed to reflect all the different kinds of rays which it receives, and so the object looks white. Any object which is black, is, on the other hand, supposed to absorb all the rays of light and to reflect none. A red object is supposed to absorb all the rays which are *not* red, while the circumstances of its reflecting the red ones is supposed to be the cause of its red appearance. So again blue and all other colours are supposed to be similarly caused by the absorption of certain rays, and the reflexion of those which seem to belong to the several objects seen by us as possessing corresponding tints. In any coloured transparent object, such as red-coloured glass, its colour is deemed to be due to the absorption by it of all the colours save the red rays, which, being unabsorbed, it transmits to the eye of the spectator.

Those peculiar conditions, activities, or what not, which exist in ordinary light and are called "colours," possess different degrees of refrangibility. This is easily shown by a simple experiment which was first tried by Sir Isaac Newton. He allowed a sunbeam to enter a dark room through a small aperture and throw a bright spot of light on a screen, placed opposite the aperture on purpose that the beam might fall upon it. He then placed a prism of glass—the opposite sides of which approached each other from above downwards at a considerable angle—in the path of the sun-beam. Thereby the beam became much refracted, so that it produced an elongated bright spot much higher up the screen than it would have fallen but for the prism. When a broad part of the prism was downwards so that its sides approached each other from below upwards, the beam was reflected downwards instead of upwards. But change of place was by no means the interesting point of the experiment. In the first place the bright spot became, as before said, elongated, and secondly it exhibited the hues of the rainbow—the red being lowest and the violet uppermost. This dispersion of the different coloured rays showed that they are bent (refracted) unequally in passing through the prism. The elongated coloured spot is called a *spectrum*. The ordinary spectrum thus formed by the sun's light— called a *solar spectrum*—may be said to be an elongated image of the sun. But spectra may be formed by light from other sources. It is always the violet rays which are the most refrangible, and the red which are the least so—always and in every medium. Nevertheless the ratio, or proportion, between the highest and lowest degrees of refrangibility is different in different media, (*e.g.*, plate glass thus differs from flint glass), and this enables the optician to construct what are called *achromatic* instruments. These are instruments purposely so arranged as to do away with the disturbing effects of the dispersion of colours which, without such aid, would take place in optical instruments, and greatly mar their utility.

With the exception of bodies which are themselves luminous, none can appear of any colour which does not exist in the light they receive. Thus if only green light be supplied to a red object, it will appear neither red nor green, but perfectly black; for the green rays will be absorbed and not reflected as such. All distinctions of colour, save differences of intensity, may be made to disappear from a room which is only supplied with light of one amount of refrangibility; but if ordinary light be admitted to play on any part of the room, the part so illuminated will immediately reappear in its natural colours. Objects seen through a coloured medium or which have had coloured light reflected on them, appear —as every one knows—to be of the colour of the light so reflected on them or of the colour of the medium through which they are viewed.

Rays which are less refrangible than the red rays cannot be recognised by the eye. If a body be raised to a temperature of 800° its rays of radiant heat will so illuminate it as to cause it to appear red hot, and at a yet higher temperature such an object may appear white. If the temperature be below 800°, however, its heat rays will not cause it to be visible.

But there are rays of light which are more refrangible than those of violet light. They cannot act on the eye so as to produce any sensation of light, though they have potent effects of another kind.

It is these highly refrangible rays which act upon photographic plates and they can produce other changes, on which account they are termed *actinic*, and sometimes *chemical* rays. Thus there are rays of very different degrees of refrangibility, only the middle series of which serve to illuminate objects and so can be recognised by our sight. The extremes of the whole known series can only be recognised in other modes. But there may be rays which, as yet, have not been recognised in any way, and which may produce effects of which we at present either know nothing, or falsely attribute to other causes.

Nevertheless not all the rays which come between the actinic and the heat rays always make themselves visible to the human eye. This has been ascertained by examining an exceedingly narrow line of light passing through a very perfect prism. The spectrum so produced appears as a band of coloured light which is continuous save that it is interrupted by very numerous dark parallel lines indicating so many rays which do not make their existence visible. The arrangement of these dark lines is different in light derived from different sources.

As has been said, the favourite hypothesis now made use of to co-ordinate the various phenomena of light, is that of regular vibrations or oscillations of ether. Whatever may be the absolute truth, there certainly is a definite periodical action or influence of some kind which takes place with amazing rapidity, but a rapidity which is different in the differently coloured rays. For the sake of simplicity of illustration, we may represent these periodical actions, or influences, as so many steps. It has been ascertained that a violet ray takes 64,631 such steps in every inch of space it traverses. Such a ray has been calculated to take about 786,000,000,000,000 steps in a second, while a red ray takes about 449,000,000,000,000 steps, and as, when unretarded by any medium, the different rays advance at the same speed, the steps of the red ray must be much longer than those of the violet one. But different rays are retarded unequally in passing through different media, and this fact explains both what is called the *interference of light* and the phenomena of *iridescence*. The steps taken by any two rays of light of the same degree of refrangibility, are equal in length. If, however, they are together to produce a more visible effect than a single ray produces, they must "keep step" —their actions must be synchronous. This shows that in each step there are two actions or influences, which are opposed to each other and which we may represent by those most generalised signs + and −.

If the + action of one ray coincides with the + of the other ray (both being of equal intensity) the visible effect will be doubled; but if the + action of one synchronises with the − action of the other, the effect of each will be neutralised and produce darkness, or a case of *interference* of light. Similarly, if white light falls upon any body the surface of which is minutely varied or which contains two or more reflecting surfaces, the rays, the steps of which are of different lengths, will be variously reflected or transmitted; and so the white light will break into colours and we shall have the phenomenon known as *iridescence*. Thus it is that, (1) mother of pearl (made up of extremely thin laminæ super-imposed), or (2) a surface marked with parallel grooves exceedingly close together, or (3) two glass plates with a very thin film of air between them, or (4) the very thin film constituting a soap bubble, will each and all appear iridescent.

The last phenomenon of light to which we deem it here necessary to refer, is that termed *polarisation*. Light coming from the sun, or from any incandescent, self-luminous body, can be reflected or refracted equally well (equally brightly) in all directions, according to the movement of rotation which may be imparted to the body which reflects or refracts it. This, however, is not the case with light which does not come from a self-luminous body but from one which merely reflects light. Such reflected light, when a movement of rotation is imparted to the body which reflects or refracts it, cannot be reflected or refracted equally well in all directions, but will be more intense when sent in some directions than in others. Such light may be said to have acquired "sides," or some property which facilitates its energy in some directions and restrains it in others. This property is called *polarity*, and (as we might now expect from what has been here noted as to the resemblance between light and heat) analogous phenomena with respect to heat,* show that heat rays have also their polarity. When we come to consider "magnetism," we shall meet with another kind of polarity, which is conspicuously manifested and very remarkable in its effects.

SOUND. — As the recurrence of certain actions, or influences, which take place in bodies, produces in us perceptions of heat or light, so the occurrence, or recurrence, of certain motions of another kind, produces in us perceptions of sound.

If two solid bodies are struck together, the shock of their contact gives rise to a sound, and certain bodies are distinguished as *sonorous* because very slight impulses will cause them to give forth very perceptible sounds, as, for example, the strings of a fiddle or the metal of a gong.

An impulse given to any body surrounded by air, is necessarily imparted by it to the immediately adjacent portion of that aëriform fluid. This, owing to the elasticity of air, rebounds after transmitting an impulse to the next portion of air and so on—the impulse being transmitted by waves through the air in all directions from the first starting-point. These aërial waves, or oscillations, may be compared with the waves which pass over the surface of a field of wheat when agitated by wind, and they thus pass along (at a temperature of 62°) at a rate of 1125 feet in a second. The air of course does not thus pass along (any more than the wheat does), but only the waves of motion traversing it. The particles of air, after each displacement, return to their former

---

* See *ante*, p. 101.

positions as the heads of the wheat do; but the regular succession of periodic motions gives the appearance of an onward wave motion of the material disturbed.

All sounds, whatever their nature, travel through the air at the same speed and pass through it in all directions simultaneously, while retaining nevertheless their distinctness—as the sounds of birds, bells, cattle and cart-wheels each remain distinct.

Since sound travels at so very slow a rate compared with light, it is easy to understand how it is we do not hear the sound of a gun till a very appreciable time after we have seen the smoke from it. Even each blow of a man beating carpets some six hundred or seven hundred yards off, will not be heard till after the eye has seen the corresponding movement.

Sound, like heat and light, diminishes in intensity with the square of the distance, and will be reflected according to the laws of equal angles. It is to this reflexion that all echoes are due, and these may be double or triple or more numerous, according to the arrangement of the surfaces on which the waves impinge.

Waves of sound which succeed each other at equal intervals and with sufficient frequency, cause us to be aware of a musical sound or note. If the waves do not succeed each other as quickly as sixteen times in a second, we have no such experience, but only a rattling sound is produced. The more rapid the succession, the higher the musical note perceived; thirty-two vibrations in a second produce almost the deepest note generally audible, and 70,000 vibrations produce the shrillest. As we saw that the human eye can only appreciate what we take to be a certain medium amount of ether vibrations, and not lower or more rapid ones, so the range of our perceptions of musical sounds is analogously limited. There are some persons who can hear the very shrill cry of the bat, but to many it is quite inaudible.

When one musical note is said to be an *octave* above another, this means that, to reveal it to us, there must be twice as many vibrations as in the case of the lower note. In notes emitted from the vibrations of strings, the longer the string the deeper the note, and to produce a note an octave higher than that produced by a string of any given length, the string must be shortened one half.

There are what are called *reed* instruments (*e.g.*, the clarionet), and in them the air blown by the player strikes deeply on a little blade placed in the mouthpiece of the instrument and causes it to vibrate, so eliciting various sounds. In a cornet-a-piston there is no artificial reed, but the player's lips are made to vibrate as if they were two reeds—one on either side of the orifice of the instrument. The vibrations of the lips are transmitted to the air within the cornet which then emits very intensified sounds.

Sounds may be intensified in various ways. They will be so if sonorous vibrations be transmitted to the walls of an empty box, and such is the action of the wooden case of the violin, without which the musical sounds of the strings would be greatly enfeebled.

The waves of air which occasion musical notes, differ greatly in length as well as in rapidity. Waves which occasion very low notes may be sixty-four feet long, whereas high ones may be less than an inch.

Differences of *timbre* are due to the relations which may exist between the main series of vibrations and secondary ones. If these secondary vibrations are synchronous with the primary ones, beauty and perfection of timbre is the result. When certain bodies are in vibration, they will elicit corresponding movements—called sympathetic vibrations—from other bodies. Thus certain wires of a piano may be made to vibrate without being touched save by the waves of air set in motion by another musical instrument emitting the notes with which such wires of the piano correspond. Even the pendulum of a clock that has stopped may be set in motion by the pendulum of another clock standing against the same wall and duly oscillating.

The vibrations accompanying sound may be perceived by other senses than the sense of hearing. Thus the vibrations of a tuning-fork are visible, and those of an organ-pipe may be very distinctly perceived by touch, as also may the oscillations of a deep-toned cord.

Air is a bad conductor of sound, which can be much better transmitted by liquids or even by solids. Water will transmit sound more than four times as fast as air, and wood or iron will carry it seventeen times faster.

ELECTRICITY. — The physical energies yet noticed, must have forced themselves on man's observation from the first, but of those which remain to be considered (apart from certain isolated phenomena) a knowledge has been acquired only during the last few centuries.

If sealing-wax be rubbed briskly for some seconds with a piece of cloth or flannel, and then be held over small fragments of torn paper within a distance of a quarter of an inch, such fragments will immediately rise and adhere to it. The same thing will occur if a glass rod be similarly rubbed with a silk handkerchief.

If a small ball, made of pith, be freely suspended by a silk thread from some supporting object, it will be attracted (like the pieces of paper) and adhere either to the sealing-wax or the glass rod. But after contact with either, it will be repelled and will diverge in an opposite direction, if that which before attracted it is made to again approach it. But though it will be thus repelled by whichever (wax or glass) was brought in contact with it, it will be attracted by the other, till it has come in contact with that other, which will then in turn repel it. Thus the pith ball may be first attracted by the wax and then repelled from it and attracted by the glass; then, in turn, it may be repelled from the glass and attracted by the wax, and so on alternately for any length of time, the friction of the flannel with wax and of the silk with the glass being again and again renewed. If two pith balls be similarly suspended side by side, then, when both have simultaneously touched either the wax or the glass, they will not only be repelled by the approach of whichever of these they may have touched, but they will also repel each other. Nevertheless this repulsion will gradually diminish, till they fall together side by side, as they were at first, when it has ceased altogether.

Thus it is evident that some peculiar influence or energy is excited by these frictions, and the force producing this energy is called *electricity*. It is also sometimes spoken of as if it were itself a substance of some kind and is often popularly called the "electric fluid."

Its real nature is as yet entirely unknown, but (like light and heat) it is well to freely make use of any hypothesis which may help us to elucidate and predict electrical phenomena without pinning our faith to the truth of any such hypothetical explanation.

There are also evidently two kinds, or states, of electricity: (1) that produced on the glass which is rubbed by silk and which kind is called *positive*, or *vitreous* electricity; the other (2), that produced on the wax by the cloth, is termed *negative*, or *resinous* electricity. Further each kind of electricity is communicable to a fresh object by contact therewith.

The experiment with the pith balls also shows that (*a*) two bodies charged with similar kinds of electricity repel each other, and (*b*) two bodies charged with different kinds of electricity attract each other.

A further examination of such bodies as we have supposed to be experimented on, will show us another most important law. If when the balls have been repelled by the approach of the wax after previous contact with it, the flannel, which has been used to rub the wax, be brought near them, they will be attracted by it, just as by a glass rod which has been rubbed with silk; while if the silk so used be brought near them, they will be again repelled. This shows that at the same time that the glass is acquiring *positive* electricity, the silk which rubs it is acquiring *negative* electricity, and that while the wax is acquiring *negative* electricity the flannel is acquiring *positive*; so that, bearing in mind the extended use* given to two mathematical signs, the electricity of both the glass and flannel may be distinguished by the sign + and that of the silk and wax may be alike denoted by the sign −. Hence we see that one kind of electricity cannot be evoked without at the very same time evoking the opposite kind of electricity somewhere else. So true is this, that if a piece of glass have a small disc of metal attached on either side of it and be suspended by a silk cord, then, if one of these discs be made positively electrical, that alone will cause

---

* See *ante*, p. 25.

the disc on the other side to become negatively electrical (and *vice versa*) without anything being directly done to the opposite side. This, as it were spontaneous, evocation of a definite and antithetical kind of electricity, is called electrical *induction*.

When two bodies in opposite electrical conditions are brought near each other they are mutually attracted, and when they approach within a certain distance (the amount of which varies with the intensity of the electrical energy excited) the electrical energy will be manifested by a flash of light accompanied by some heat and a sound. Then occurs what, on a small scale, is called an electric spark, and on a large scale is known as a flash of lightning. When this has taken place sufficiently, the opposite surfaces will be found to be no longer in opposite electrical states; the "discharge," as it is called, will have neutralised their opposition, and both will have returned to their normal and unexcited condition.

But a phenomenon much more conspicuous and familiar than that of induction, is what is known as "*conduction*." The pith balls and the glass disc were supposed to be suspended by silk, because, while so suspended, they can keep the electricity they acquire. If they were suspended by a metal wire, they would not keep it at all, for it would instantaneously run away through such a channel. The metal wire takes, or "*conducts*," it away with extreme rapidity and facility, metals being extremely good *conductors*. Thus bodies may be arranged in two opposite classes: "conductors" and "non-conductors." Silk, glass, resin, wax, porcelain, and india-rubber are all non-conductors, and woollen material, dry wood, and leather conduct badly. Not only the metals, however, but all objects containing much moisture, such as the bodies of animals and plants, are good conductors.

Non-conductors are also called *insulators*, because they serve to insulate electrified bodies, as the silk thread insulates the pith balls and so keeps their electricity from passing away. Air, especially dry air, is a bad conductor, and this is why electricity, when the tension becomes too strong, flashes through it as a spark, instead of being quietly conducted by it from one electrified body in one state to another body in the opposite electrical condition. But air is not an absolute non-conductor, and hence it is that by degrees the two similarly electrified pith balls slowly part with their special electricity and therewith—repulsion ceasing—fall together.

Bodies in which electricity can be easily excited (*e.g.*, glass and sealing-wax) are in general the worst conductors, but it seems that it can be excited in all bodies by friction and, as we shall see later, by other means also; only in very many cases it runs away by conduction as quickly as it is generated. But whenever so good a conductor as a metal rod is fitted with a glass handle and so insulated, it can be excited by rubbing and will retain its electricity.

The rapidity with which the electrical energy travels is enormous, the velocity of electrical disturbance (or energy) being the same as that of light.*

The facts of conduction and insulation enable us to accumulate electrical energy. The most familiar form of its accumulation is in what is known as a *Leyden Jar*. This is a glass jar, or wide bottle, all but the upper part of which is coated both inside and outside with tinfoil. The glass keeps each of the two layers of tinfoil insulated from the other, and the inner layer is entirely insulated. The neck of the jar is then closed with a cork, through

---

* See *ante*, p. 103.

which passes a piece of metal (wire or a chain) which internally makes contact with the inner layer of foil, while externally it terminates in a rounded knob projecting freely from above the cork of the jar.

Then electricity is produced by that familiar instrument called an *electrical machine*, which consists of a disc, or a cylinder, of glass, capable of being rapidly turned between, and so becoming rubbed by, two cushions; such action (as in the experiment with the silk and the glass rod) developing electricity on the glass. This is collected, as it is generated, by certain metal parts of the machine, which are insulated on glass legs and which end in a projecting knob, whence in very powerful machines sparks may sometimes be obtained half a yard long and strong enough to knock a man down.

Electricity having been thus generated, it is carried by a conductor to the knob on the Leyden jar, whence it passes, through the wire, or chain, to the inner coating of foil * which becomes charged with (positive) electricity when the outer coating of the jar also becomes, by induction, simultaneously charged with electricity of the opposite kind—in this case negative. Then if a piece of metal (with a glass handle for safety from shock) be made to touch any part of the outer layer of foil with one end, while its other end is applied to the knob of the jar, an instantaneous discharge of electricity takes place, and the two electricities neutralise each other and disappear. Before contact is effected, a brilliant spark with a more or less loud report, will pass between the

---

* The electricity is thus by no means stored in the bottle, as would be the case with a liquid. That, when charged, the particles which compose the glass assume a peculiar arrangement is a favourite hypothesis.

jar's knob to the conducting piece of metal. A round form is given to the knob because it is much more easy to retain electricity in a body of that shape. In projecting edges and points, the energy becomes so concentrated and intense, that it cannot well be confined in bodies so shaped.

When a metallic circuit is complete, a current of electricity may readily be made to traverse it, but the current ceases immediately the circuit is broken. Electricity is conducted along copper wire with extreme rapidity, as before stated, and currents may be exceedingly powerful. If the metallic circuit through which such a current passes be of unequal capacity, then its thinnest part may become, and be maintained at, a red or even a white heat, reminding us of the accelerated flow of a river where its banks approach each other and the stream is narrowed.

Currents of electricity have very wonderful and complex effects to which it is impossible here to allude more than distantly, the reader being referred, for all but the very elements of the science, to special works on *Electro-dynamics* and on the relations of electricity to heat, or *thermo-electricity*.

But the effects of currents of electricity and certain modes of their generation, cannot be understood without some elementary notions of magnetism and chemical energy, so that what further remains for us to say about electricity will be said when we come to speak of those energies.* It may, however, be here briefly noted that, whereas bodies in similar electrical conditions repel, and in opposite electrical conditions attract, each other, currents flowing in the same direction

---

* See *post*, pp. 127, 133 and 134.

attract, while those flowing in opposite directions repel each other.

Heat and electricity have very definite relations, and a multitude of more or less recent investigations have enriched the thermal section of electrical science. Here it must suffice to say that if certain metals are heated unequally, an electrical current from the hotter portion of the metal to the colder, and from the colder to the hotter (if there is an uninterrupted metallic circuit), will be set up. A current may also be set up in a circuit of conducting material by the application of heat to one part of it, provided the heat be so applied that it does not diminish symmetrically on each side of the portion most heated. Thus the motion of heat produces electricity, and (as we have seen with respect to electric currents) the motion of electricity, so to speak, produces heat and may produce light.

Some further effects of electricity and its relations with chemical energy, will be noticed towards the close of this chapter.

MAGNETISM.—The common horse-shoe magnet, with which many children are familiar, consists of a bar of steel bent so that its two ends come near each other. It has a slight action on certain substances, probably some action (to us as yet imperceptible), on all bodies in its vicinity, but a great and conspicuous effect on steel and iron. Needles, iron filings and other small iron bodies are attracted to, and will rise and join, such a magnet when its ends are held near them. If a variety of small bodies—sand, cinders, small fragments of wood, tiny bits of silver and iron filings—be all mixed together, such a magnet will readily separate out the iron particles, which will cling in bunches to its two ends. Even if the iron filings be spread upon a piece of paper or a thin sheet of glass, and the magnet be moved above on the under surface of the paper or glass, the iron filings will follow the magnet so that its influence is clearly transmitted through such bodies.

The magnets first known consisted of a certain mineral called "*loadstone*," which is especially abundant in Sweden and in Asia Minor. Loadstones attract iron and steel, while they themselves consist of an iron ore— that is, of iron, in union with certain other substances, or at least, resolvable into such.

Artificial steel magnets can be made from loadstone by rubbing them against that mineral, but they are now also made in other ways. A piece of pure iron, by contact with a natural or artificial magnet, will itself become magnetised and attract iron filings, steel needles, &c., so long as it remains in such contact, but it does not permanently retain its magnetic power. Such is not the case with steel, which by contact, or rubbing, will become itself permanently magnetised.

It is the ends of the magnets which mainly attract, the bent part produces hardly any effect. But the actions of the two ends are never alike.

If a magnetised steel needle be freely suspended and allowed to oscillate till it becomes motionless, then if one end, A, of a horse-shoe magnet be brought near to one, X, of the two ends of the needle, the needle will be attracted and rotate so as to bring that end, X, as close as possible to, or in contact with, the end, A, of the magnet. Then, if the other end, B, of the magnet be brought near that same end, X, of the needle, instead of being attracted, it will thereby be repelled. Yet that same end, B, of the magnet will attract the opposite end (or opposite pole Y) of the needle, which will, on the contrary, be repelled by the other pole, or end, X, which was first presented to the needle.

It would thus seem, judging from the phenomena of the pith balls before noted,* that there are two kinds of magnetism, as there are two kinds of electricity, and this may be made certain by the following simple experiment. Take two needles, the ends of which we may distinguish as A and B. Magnetise them by rubbing several times from the middle of each towards the points, but towards the points A and A with one end of the horse-shoe magnet, and towards the points B and B by its other end. Then evidently the A poles of the two needles will have gained one kind of magnetism, and their B poles the other kind. Thereupon it will be found that both their A and B poles repel each other, while the A pole of each will attract the B pole of the other needle and *vice versâ*. Therefore of the two kinds of magnetism—like the two kinds of electricity—it may be said that like repels like but attracts unlike.

We have before spoken † of the polarity of light, but polarity is thus especially manifest in magnets and by means of magnetism. The polarity of magnetism is like that of electricity, in that one cannot exist without the existence of the other in its neighbourhood. But two bodies can be in different electrical conditions, each having only one kind of electricity, as *e.g.*, in the silk and the glass rod. Such, however, is not the case with magnetism, since every magnetic body must have, in itself, both of the magnetic polarities. However small the fragments into which a magnet may be broken, each fragment will always possess both kinds of magnetism and have two poles and not one only.

---

* See *ante*, p. 119. † See *ante*, p. 113.

This wonderful form of energy, as before noted, is not confined to iron or steel, and no substance is completely indifferent to it. Bodies which only manifest its influence by being attracted, have been named *magnetics*. Those, the far larger class, which are only repelled—repelled by both poles—are called *diamagnetics*. The wonderful polarity of the magnet will even modify the polarity of light; but such matters are beyond the range of those elementary notions of science with which alone this work has to do.

We have before spoken of the relations between magnetism and electricity, and they are indeed very close relations. An electric current passing near iron or steel tends to make it magnetic, while the movements of a magnet tend to produce electric currents. The rotation of a magnet on its axis will produce electric currents through it, if the universally necessary condition of a current, a complete circuit, exists. On the other hand, the passage of electric currents around an axis will make the axis it surrounds magnetic.* Let a rod of iron have an insulated copper wire wound a number of times about it. Then so soon as a current of electricity is made to pass through the wire, so soon will the rod of iron become a magnet—an electro-magnet—but its magnetic energy will cease the moment the current of electricity is discontinued. The iron may be bent as a horse-shoe and then if, through a copper wire copiously investing it, a strong current of electricity be made to pass, the iron horse-shoe will, as long as the current so passes, be a very powerful electro-magnet, and sustain a great weight of iron, which, however, will instantly fall away from it, the moment the current ceases.

---

* For a very important consequence of this, see *post*, p. 171.

CHEMICAL ENERGY.*—As is manifest even to children, a number of different substances exist, such as sand, stones of different sorts, diamonds, glass, matter known as "salts" of various kinds, silver, gold, iron, and other metals, as well as water, quicksilver, air, the gas which we burn, &c. &c.

Most of these bodies can be "resolved," by one or another process, into other substances which appear to have composed them, and this appearance is greatly strengthened by the fact that the substances so resolved can often be reproduced by the bringing together, under certain conditions, of the matters into which they were previously resolved. At first it would seem as if such substances have been merely mixed, and that their component parts can be disentangled and then mixed again. Such, however, does not seem to be the case.

We all know how readily iron rusts when exposed to the air. But the rust of iron is not iron, nor can we affirm it to be iron mechanically mixed with something else so as to be able, by another mechanical process, to be separated from it again. Iron is one substance, iron rust is another and diverse substance. It is a thing of a different nature, with a number of qualities quite distinct from those of anything which may, with the iron, have contributed to form it—to form the rust. And another substance has so contributed—a gas contained in the air and known as *oxygen*. This gas, by acting on the iron at its surface, coalesces with it to form the new substance, "iron rust," or what chemists call *oxide of iron*, and the process itself is called *oxidation*. By heating it in a

---

* Some additional information about chemistry will be found in the next chapter (pp. 138 to 145), wherein various properties of minerals are considered.

certain manner, the rust can again be resolved into oxygen and iron, but no mechanical process will separate them from each other. It is the same gas which, by its chemical action on other substances, causes flame, and all our fires are manifestations of such chemical energy.

When we mix together, in water, the two finely divided bodies which are often given to us as "effervescing powders," these two bodies do not merely become liquid and mix. They each become resolved into other substances which then partly unite together (in new combinations to form a new liquid body of a different nature) and partly give rise to an aëriform body—a gas —which escapes from the liquid in bubbles, so causing the effervescence. This is very different indeed from a mechanical mixture such, for example, as when we mix together two dry powders of different colours.

The gas, oxygen, which, by uniting with iron, forms "rust," exists in the atmosphere mechanically mixed with the air's other gaseous components. Water, however, can be resolved into that same gas and another gas called *hydrogen*, and these are not merely mechanically blended. If these two gases be mixed, in due proportion, within a carefully closed glass vessel, and an electric spark be made to pass through the mixture, the two gases will entirely disappear and water will be found to exist in their stead.

Chemical energy will therefore actually transform two or more substances into two or more other substances presenting characters which seem to indicate a quite different nature, with quite different properties. Every substance possesses its own chemical energies, either dormant or potential* (as in effervescing powders before they

---

* See *ante*, p. 85.

are mixed in water), or in an active, kinetic state (as when they rush together in their effervescence), and then as such chemical energy ceases and disappears, it gives rise to, and is replaced by, a definite amount of heat. Chemical combination therefore appears to be an altogether different thing from a mere mixing or blending of substances, each of which retains its own essential characteristics. Different substances may be merely mixed in any proportion, as we may colour water the more by putting in it more indigo, or as we may mix sand, or sulphur, and iron filings in any quantity we like. All the masses thus produced consist of minute separate particles, each of which retains its own properties. But if we mix together sulphur and iron filings and pour some warm water on the mixture, we shall then have a manifestation of chemical energy resulting in the production of something altogether different from the bodies which before existed, more or less of which will be thus made to disappear. The mixed mass becomes hotter, swells and assumes a blackish colour, and the new body which thus makes its appearance is what chemists call *sulphide of iron*. Such is the difference between a mechanical mixture and a chemical combination.

A careful study of the process of resolving some substances into others (called *analysis*), and of the opposite process of producing new substances by the junction of others (called *synthesis*), shows us that chemical transformations take place in an exactly definite manner as estimated by weight.

Thus, in forming sulphide of iron, four equivalent units of sulphur and seven of iron will disappear and eleven units of sulphide of iron will be produced. If there be more than this of either material, then such odd quantity left over will remain unchanged, alongside of the sulphide produced.

There are certain substances which are termed "elementary" and chemical "*elements*," because they do not seem capable of being resolved into other substances, and some of these elements have an overpowering attraction for each other. We have already noted that which exists between iron and oxygen, which results in the "rust" or "oxide of iron." Similarly, a metal, *calcium*, and oxygen will rush together and produce a rust, or *oxide of calcium*, which is *lime*. Oxygen and another metal, termed *silicon*, will unite to produce a rust known as *oxide of silicon*, which is *flint*, or *silica*. Similarly one component of clay, namely *alumina*, is an oxide of yet another metal—*aluminium*.

Not only elements, but various compound substances (so called because capable of being resolved into others), when placed in close proximity to certain other substances, will undergo a spontaneous transformation and, as it were, exchange partners; and this is often facilitated by their being dissolved in water. Thus in the case of the common effervescing powders before referred to, we may have one powder consisting of carbonate of soda (which can be resolved into carbonic acid and soda)* and the other powder formed of the acid of lemons, or *citric* acid. When these are simultaneously dissolved, the citric acid will seize upon and unite with the soda, while the carbonic acid of the carbonate of soda is set free and escapes in a gaseous form in the bubbles of the effervescence.

It may be that a substance resolvable into two elements, may be robbed of one of its elements by a third element placed in its vicinity, so producing a new and

---

* See *post*, p. 140.

different binary compound. Thus water, as we have seen, consists of two elements, oxygen and hydrogen, chemically united, and they are in the proportion of two volumes of hydrogen for one volume of oxygen. Now a very light metal, known as potassium, and oxygen, have an extreme chemical affinity for each other. This is so great that if a piece of potassium be thrown into water the chemical energy thus developed is of such intensity that heat and light are produced in the water, by the energetic junction of some of the oxygen of the water with the metal, which burns (so forming potash or *oxide of potassium*), while such of the hydrogen of the water as is thus abandoned by its oxygen, being set free, also burns in uniting energetically with the oxygen in the air, so forming aqueous vapour. Another such reciprocal interchange of elements will take place if we put together the two substances respectively known as "*nitrate of silver*" and "*hydrochloric acid.*" Nitrate of silver is resolvable into silver, nitrogen and oxygen, there being, by weight, three times as much oxygen as either of the others. The hydrochloric acid consists of the gases *chlorine* and *hydrogen*. When these two substances are placed in proximity, the chlorine will leave the acid to unite with the silver of the nitrate of silver, and so produce what is called *chloride of silver*, while the hydrogen of the acid unites with the nitrogen and oxygen of the nitrate of silver and so forms *nitric** acid. It is practically most useful, in the study of chemistry, to regard each substance as made up of minute particles called "*molecules*," and these again of still more minute particles termed "*atoms*," combined in definite proportions

---

* As to "nitrates," "carbonates," "sulphates," "phosphates," and for other examples of chemical change, see *post*, pp. 139-142.

—as estimated by weight—according to what is known as the "*atomic theory.*" A whole series of symbols has been devised to conveniently denote such conditions; thus water is imagined to be made up of molecules, each of which is composed of two atoms of hydrogen and one atom of oxygen. Water therefore is represented by the symbol H_2O, and H_2SO_4 stands for *sulphuric acid*—S signifying sulphur. Similarly, a multitude of analogous symbols serve to denote a multitude of different substances. This system is of enormous practical utility, and demands the most careful study, though the absolute constitution of matter remains unknown to us. For further information about the laws concerning the proportions in which different elements and chemical substances combine, and the symbols made use of to denote such proportions, the reader must have recourse to treatises on chemistry.

Chemical energy is very closely related to the other physical energies. It not only occasions warmth but may be greatly facilitated thereby, while the heat and light it may occasion can be made manifest by acting as just described, with potassium. It is, however, also exemplified by every fire and every candle which we see burning, since every such phenomenon is both a result and a sign of the exertion of chemical energy. Chemical energy is most intimately related to electricity. Indeed the resolution, by chemical energy, of different substances is always accompanied by some electrical change.

An electric current, on the other hand, can produce chemical changes in substances through which it passes, and so bring about changes not as yet otherwise producible, as we have just seen with regard to a spark traversing a mixture of oxygen and hydrogen and so evolving water. There are certain chemical substances termed *acids* (*e.g.*, sulphuric and nitric acid), and others termed *alkaline* substances, each of which latter is also called a *base*. Bodies composed of an acid and a base are termed "*salts*." Now if the two poles (the ends of the wires) of an electric pile or battery be placed in water so that the electricity may be conducted through it from one pole to the other, and any "salts" be dissolved in that water, we shall find the solution breaks up, the acid matter going to the positive pole, and the base, or alkaline constituent, to the negative pole. This strikingly demonstrates the close connection which exists between electrical and chemical energy. The quantity of electricity generated by chemical changes is often enormous. It has been said that the resolution of a grain of water will generate a quantity equal to a flash of lightning. As in the generation of heat, light and magnetism by chemical energy, so also in its generation of electricity, the quantity generated bears a constant and exact relation to the amount of chemical energy expended.

The mode in which electricity is most commonly and effectively generated by chemical energy and a powerful electric current induced, is by what are called *electric batteries* and *piles*. The simplest and most primitive form of the electric pile consists of a series of discs of zinc and copper, separated by discs of cloth soaked with vinegar, and with two wires fastened to the series. One of these wires must touch a disc of copper at one end of the pile and the other must be in contact with a disc of zinc at the other end of the pile, while each wire projects freely at its other end. The chemical energy excited by the contact of the substances in the pile gives rise to electricity, and therefore to both its so-called kinds. The positive kind goes exclusively to the wire which touches the copper, while that from the zinc receives the negative electricity. If now the two free ends of the wires be joined, a current will be immediately established, and, by making such a current pass round iron, magnetism is, as before stated, necessarily and at once induced.

By means of small porcelain jars, conjoined in various series, and each supplied with an acid fluid in which portions of copper and zinc are suspended, powerful generators of electricity (batteries) are formed, the wires from either extremity of which (bearing, to speak popularly, different electricities) will, by their approximation, chemically resolve many substances and will emit small sparks at each contact. When these currents are made to pass between small pieces, or delicate filaments, of pure charcoal attached one to the end of each wire, the brilliance we know as "the electric light" will be generated. Such currents also serve to work the electric telegraph, the telephone, and the microphone, which here we can do no more than name.

Chemical changes between certain bodies may be induced by the mere proximity of similar changes going on in other bodies, and examples of this process are met with in photography.

When chemical energy takes place with great intensity, it is generally accompanied by light and heat and the quantitative relations before referred to[*] as existing between them, and between them and magnetism and electricity, reveal to us a profound similarity between all known physical energies.

As to what chemical energy is in itself, we know no more than we do with respect to the other forms of energy. It may be that there is in Nature either one

---

[*] See *ante*, p. 102.

unimaginable form of energy which manifests itself according to circumstances—as heat, light, electricity, magnetism and chemical energy; or else that there are fundamentally different forms of force which have between them relations of quantitative equivalence as they act. The former is the view now popular. These, however, are questions so remote from the mere elements of science, that we will no further deal with them here, but proceed to consider that real world wherein all these physical energies play their several parts.

CHAPTER V

THE NON-LIVING WORLD

HAVING now made some acquaintance with the laws which regulate the stability and movements of bodies, and with the various forces which may energise in them, we may next proceed to survey the actual world about us. We have to study the nature, structure and properties of the parts which actually compose it, and the various ways in which, as a whole, it is modified by the physical forces which act upon it, making abstraction, however, of the phenomena of life. We assume that the student knows the earth to be globular, with a north and south pole equidistant from the equator; also that its surface is described by means of imaginary circles, parallel with the equator, marking degrees of *latitude*, and by others which, at right angles with the former, pass through the poles and serve to indicate degrees of *longitude*. The world is everywhere surrounded by an aëriform mass, the *atmosphere*, while the greater part of its surface is covered, more or less deeply, by water. Its solid mass is, as every one knows, composed of a variety of matters, such as different kinds of earth, some being clay, some sand, &c., with many stones scattered through and over its soil, while large tracts are composed of rocks. These rocks may be sandstones, slate, granite, limestone or chalk, &c., and the rocks may contain metals, metallic ores or crystals. The whole mass of different substances which thus make up the solid substance of the globe are termed "minerals."

As was pointed out in the last chapter, most of these bodies can be resolved into other substances, and ultimately into "elements"—so called because they have not been found capable of further chemical analysis. There are about seventy substances which are thus provisionally regarded as ultimate, and which, by most varied and different degrees of chemical synthesis, compose all those matters whereof the world consists. But however varied may be the degrees of synthesis produced, the elements, as before said,* are always combined in each kind of substance, in one exactly definite manner, as estimated by weight. Of the various elements, some, at what to us are normal, moderate temperatures, are aëriform. Such are the gases before spoken of as *oxygen* or *hydrogen*,† and also the gases *nitrogen* and *chlorine*, with various others. All the metals are elements and are normally solid, though mercury, as we know, is liquid. Amongst the metals are *calcium, silicon, aluminium, potassium, sodium, magnesium,* and *arsenic*. Other solid elements are *carbon* (or pure charcoal), *sulphur, phosphorus,* and *iodine*.

Oxygen at all ordinary temperatures is a colourless gas, but it has lately been changed, by reducing it to an extremely low temperature, to the condition of a blue liquid. As we have seen, it has a great tendency to unite itself with many other substances. When it unites violently, the substance it unites with "burns." A general process of union, such as the rusting of iron, may be therefore called a slow combustion. Anything which burns in the air, burns with far greater intensity and brilliancy when plunged in oxygen. Nevertheless, though

---

* See *ante*, p. 133. † See *ante*, p. 129.

the greatest agent in, and supporter of, combustion, oxygen itself is incombustible in air. Various of its combinations or "oxides" have been already noted.*

Hydrogen does burn in air, though it cannot aid combustion. It is the lightest substance known, and forms the chemical standard or unit of weight, and is very widely diffused. Water can be resolved † into twice as much hydrogen (estimated by volume) as oxygen, and—as was stated in the last chapter—into one part of hydrogen to eight of oxygen as estimated by weight. Water may be called an oxide of hydrogen, and hydrogen may be regarded as an aëriform metal.

Nitrogen differs greatly from oxygen, save that, like oxygen, it is colourless. It is extremely indisposed to unite with other elements, and, so far from promoting combustion, it stops it, extinguishing a flame plunged into it. It is remarkable also for the extreme instability of the compounds of which it forms a part, such as gunpowder, gun-cotton, nitro-glycerine, and iodide, sulphide and chloride of nitrogen. These constitute a series of substances successively exploding with greater and greater violence and readiness. Nitrogen, nevertheless, is itself incombustible in air. In conjunction with oxygen it forms nitric acid, which, when united with other substances, produces what are called "*nitrates.*"

One notable substance that is resolvable into nitrogen and hydrogen, is *ammonia*, which, under ordinary circumstances, is a gaseous as well as alkaline ‡ substance. It is colourless but very pungent, and dissolves in water with extreme rapidity, that liquid at 50° Fahr. condensing 670 times its own volume of ammonia.

---

* See *ante*, p. 131. † See *ante*, p. 132.
‡ See *ante*, p. 134.

Chlorine is a gas of a green colour, and can be compressed into a yellow, limpid liquid when subjected to a pressure about four times that of the atmosphere, but rapidly becomes again gaseous when such pressure is removed. Like oxygen, it is a supporter of combustion, and powdered-antimony when thrown into it burns spontaneously. It possesses a powerful and peculiar odour. Many of its compounds are termed *chlorides*, and, united with the metal sodium, it forms *chloride of sodium*, or "common salt."

Carbon is an element which remains solid even at the highest temperatures yet applied to it, but it is rarely found pure. As such it may exist in one of three different conditions—(1) as pure charcoal; (2) as black-lead or graphite; and (3) as the diamond. It is very abundant united with oxygen. Such oxide of carbon, or carbon "rust," is a gas at all ordinary temperatures and pressures, though by extreme pressure it has been made liquid and also solid. It is commonly known as "carbonic acid," and since it is formed by the union of charcoal with oxygen (six parts, by weight, of carbon to sixteen parts of oxygen), it is given off abundantly where coal fires are burned. It is a colourless gas with little, if any, odour, and a burning candle plunged into it becomes extinguished. United with a variety of other substances it produces what are known as *carbonates*, such as *carbonate of lime* and *carbonate of soda*. Four equivalents of carbon, four of oxygen, and two of hydrogen, constitute *citric acid*, so commonly used for effervescing drinks. This acid will unite with other substances, such as ammonia, potash, soda, and lime, forming a *citrate* of each respectively.

Sulphur.—This yellow elementary mineral, commonly called "*brimstone*," is normally a solid; but a small heat will render it liquid, and a little more will make it aëriform. Then, if a cold surface be brought near, it will again become solid, being deposited on that surface in the form of minute grains, known as "flour of sulphur." But besides its liquid and gaseous states, sulphur may exist in more than one solid condition, and may be made to pass alternately backwards and forwards, from one solid condition to the other, by means of slight changes of temperature. One of these is known as crystalline sulphur, while the other is non-crystalline.* Substances which crystallise in two different forms, without change of nature, are called *dimorphic;* and those which, like carbon, can exist in more than two, are termed *polymorphic.* The same body may possess distinct properties, and any one such state contrasted with another, is said to be an "allotropic" state. Thus, we say crystalline sulphur can exist in an "allotropic" state which is not crystalline. Sulphur forms some very notable substances with the aid of other elements. Thus, *sulphuric acid*, or "oil of vitriol," is an oleaginous liquid, formed by three equivalents of oxygen, one of sulphur, and one of water. Sulphuric acid forms, with various other substances, certain matters termed *sulphates*, as sulphate of copper, sulphate of magnesia, sulphate of lime, &c. It takes away the citric acid from citrate of lime, seizing on the latter and forming sulphate of lime, the citric acid being thus left free. The termination "*ic*" signifies that the body so distinguished has more oxygen than one distinguished by the termination "*ous*"—as sulphur*ic* acid has more than sulphur*ous* acid has. *Sulphurous acid* is formed by two equivalents of oxygen and one of sulphur. It produces that powerful, suffocating

---

* As to what a crystal is, see *post*, p. 143.

odour we experience when sulphur is burnt; various bodies formed by it in conjunction with certain other matters, are termed *sulphides*.

Phosphorus can, like sulphur, exist in two distinct solid states. One of these also is crystalline; when in its not crystalline state, it is said to be in an *amorphous* condition. It has a great affinity for oxygen, and readily bursts into flame, while ordinarily it is in a state of slow combustion which makes it luminous in the dark. In union with oxygen it forms phosphorus and phosphoric acid, the latter containing, of course, the greater proportion of oxygen. Phosphoric acid when united with other substances forms what are called *phosphates*, as, *e.g.*, *phosphate of lime*.

Iodine exists in sea-water. When pure it is a soft, opaque, crystalline solid, of a bluish black colour and with a metallic lustre. When moderately heated it becomes a violet coloured vapour, which solidifies again in crystals. It has a strong, disagreeable odour and taste, and gives an intense blue colour to a solution of starch. It unites with metals, forming what are called *Iodides*.

Of the metals, gold and silver do not rust (oxidise) by exposure to the air, and they, with platinum, mercury and copper, are often found pure, or in their "native" state, as it is called.

It is just the reverse with the metals potassium,* calcium, aluminium, and also *sodium*, the oxide of which is *soda*. *Magnesium* oxidises as *magnesia*. *Silicon*, or *silicium*, unites with oxygen, as before said,† to form silica—*i.e.*, *silicic acid*, and the products of this acid with other bodies are termed *silicates*. Slate, much of what

---

* See *ante*, p. 132. † See *ante*, p. 131.

we call clay, with granite, quartz and various other stones, are formed of " silicates."

Arsenic is a metal of a steel-gray colour and considerable brilliancy, and forms with oxygen *arsenious* and *arsenic* acid.

Iron is the most abundant and widely diffused of all metals, but is rarely, if ever, met with pure. It is found in various combinations, or "*iron ores.*" The natural magnet, before spoken of,* is such an ore in the form of a crystalline oxide. Carbonate of iron mixed with various proportions of earthy matters forms alone one-third of the iron ore of Great Britain. Iron and sulphur, or *iron pyrites*, also exists in enormous quantity. There are also chlorides of iron. Steel consists of iron united with carbon.

We have spoken of certain bodies being "crystalline," which means that they are made up of crystals, large and small. Now a *crystal* is a solid mineral substance of a definite geometrical figure, being bounded by surfaces, or faces, which meet so as to form sharp edges and angles. The angles formed by these faces are characteristic of different crystalline substances, though there is no constancy as to the size of the crystals, or the proportionate size of their several faces. Snow is a very familiar example of a crystalline body, and is one which can only exist as such at a low temperature. If a crystal be suspended in water which holds in solution as much as it can contain of the same material as that whereof the crystal is composed, then if the liquid be evaporated, fresh solid material may be deposited, from the liquid, upon the surface of the crystal, which will thus increase in size. If a crystal, so suspended, be

---

* See *ante*, p. 125.

mutilated by having one of its solid angles removed, such injury will be repaired by the deposition of fresh material from the liquid.

Crystals also possess the power of thus resuming growth after interruption; and there appears to be no limit to the time after which the resumption of growth may take place. A crystal may also undergo great internal changes, and may be almost entirely disintegrated, yet if a small portion remains, it may grow and perfect itself again. Two crystals of different substances may grow so as to become almost inextricably intermixed, each of them preserving its individuality and growing according to its own laws all the time. Crystals may shoot out in an arborescent manner, as we often see in the "frost" (which consists of crystals of ice) on a window-pane.

Stones, rocks, and other substances are said to be "crystalline" when they are formed of minute crystals aggregated together—as is the case with marble and (as before noted) with one state of sulphur. Other minerals may be of similar chemical composition, but not formed of minute crystals—as, *e.g.*, chalk, and one of the solid states of sulphur.* The formation and growth of crystals evidently takes place by most minute particles, answering to the "molecules" spoken of in the last chapter, each molecule being composed of "atoms," according to the "atomic theory." It is supposed that these atoms are persistent, indestructible, and indivisible, as well as unchangeable both in weight and volume. It is also supposed that they are separated by interstitial spaces, void save that they are occupied by ether. The increase or decrease of these spaces is thus

---

* See *ante*, p. 141.

deemed to be the one explanation of the augmentation and diminution of the bulk of bodies. Finally the atoms of the chemical elements are supposed to be of determinate specific gravities. These hypotheses are of the greatest value for investigating, predicting, and producing chemical changes of analysis and synthesis. There is no need, however, to esteem them as more than working hypotheses, still less to deem them a sufficient and exhaustive explanation of the real nature of bodies. Indeed there are various considerations which at present absolutely forbid us so to regard them.

We have seen how various substances, such as carbon, sulphur, and phosphorus, may exist in two or more different states, but there are certain others which can be made to alternate between two conditions neither of which is truly crystalline though one of them is allied to the latter state. Thus the same chemical substance may sometimes exist in a state which is termed *crystalloid* and at other times in the state of what is called a *colloid*.

Substances which are in a "colloidal condition"—*i.e.*, are "colloids"—are jelly-like, and insoluble in water. They readily absorb water through their substance and swell, while they will also readily yield it up again by evaporation.

Crystalloids are not merely the reverse of all this, but are specially remarkable for their diffusibility; while colloids can hardly diffuse themselves at all through the substance of other colloids.

Substances can often be made to pass from the crystalloid to the colloidal condition by adding a minute quantity of some substance, such as an alkaline carbonate. As an example of a substance which will alternately exist in these two states, may be mentioned that known as peroxide of iron.

K

There is also a peculiar action of liquids which may here be mentioned. If two liquids of different densities are placed within a vessel so that they are separated by a median porous partition, then a portion of each liquid will pass through the partition, but more of the less dense liquid will pass through it than of the other. The consequence is that if the level of the two liquids be at first the same on each side of the partition, then the level of the denser liquid will rise, while that of the less dense liquid will sink. This process of fluid transference is called "*osmosis*," and it is facilitated if the partition be itself a colloidal substance.

We must now return from this digression (which has arisen from what it was necessary to say concerning crystals) and consider a little further what are the solid, liquid, and aëriform bodies which compose this earth.

Limestone, marble and chalk, all consist of carbonate of lime* and have been produced by the chemical union of lime and carbonic acid. If sulphuric acid be poured on any of the three substances, or if small pieces of them be placed in a solution of sulphuric acid, bubbles will be given off—there will be effervescence. This is due to the acids changing places. The sulphuric acid unites with the lime, or "base,"† and the carbonic acid, which is normally a gas, is set free—hence the effervescence. Other acids—*e.g.*, strong vinegar—will produce a similar effect. Soil, stones, and rocks which can be thus acted on are called *calcareous*. Soil, stones, and rocks, where flint or silica plays the part which lime plays in the calcareous rocks, are termed *silicious*. Such are alabaster, slate, sand and sandstone rocks. Acids have no effect upon them. Silicious crystals are extremely hard,

---

* See *ante*, p. 140. † See *ante*, p. 134.

and *quartz,* or "rock crystal," is one of the commonest of them. Rarer ones are rubies, emeralds, sapphires, topazes, amethysts, opal and the substance hydrophane, before referred to.* The most brilliant of crystals, diamonds, are, as before said,† formed of carbon, and, of course, are neither silicious nor calcareous. Great crystalline masses of rock-salt are found in many places.

Granite consists of an accumulation of crystals of three kinds intermixed—namely, quartz, and two silicates of alumina: mica and feldspar. *Gneiss* is a highly crystalline rock, composed of a feldspar with mica and quartz.

Porphyry and *basalt* are allied to granite. It has been found by experiment that such rocks can be rendered liquid at very high temperatures, and liquid rocks of such kinds exist naturally in the form of the lava‡ which is emitted from burning mountains or volcanoes.

Our *atmosphere*, the aëriform envelope of the earth, or air, is not like water, a substance resolvable into gases, chemically combined, but is a mixture of gases and vapours. About one-fifth part of it consists of oxygen and almost all the rest of nitrogen. Carbonic acid is always present in small but varying quantities, as a general rule about five volumes of it to 10,000 volumes of air.

There is also some ammonia and a certain quantity of the vapour of water. The amount of this aqueous vapour, however, varies greatly, as it must do from what we have already seen § respecting the conversion, by heat, of liquid water into vapour or steam, and its condensation, at a lower temperature, into its liquid condition once more.

---

* See *ante*, p. 103.
† See *ante*, p. 140.
‡ See *post*, p. 163.
§ See *ante*, pp. 86 and 87.

Air has lately been reduced, by Professor Dewar, to a liquid less blue than liquid oxygen.

The hotter the air, the greater the amount of aqueous vapour it can contain, and *vice versâ*. The maximum quantity which air contains at 50° F. is about $\frac{1}{120}$ of its weight, but at 82° air will contain $\frac{1}{43}$ of its weight.

If air, containing much aqueous vapour, be suddenly cooled, the latter is thereby forced to condense itself into particles of liquid water, which may appear as mist, cloud, rain, or dew. A familiar example of such condensation may be seen when a decanter of iced water is brought into a hot room; then aqueous vapour will immediately condense upon its exterior.

Dew is occasioned by the radiation * of heat from the earth's surface which, when the sky is clear, *i.e.*, when there are no clouds to reflect it back again, rapidly cools that surface, and so forces the air immediately in contact with it, to condense its vapour and part with it in the form of dew. Dew does not fall, but, as it were, grows upon the surface it coats, just as on the surface of the decanter of iced water above mentioned. Hoar frost is vapour frozen into crystals of ice.

Since the warmer the air, the greater the amount of aqueous vapour it may, but by no means must, contain, it follows that the greater amount of cooling air can stand without shedding its dew, the drier it must be— the less the proportion of aqueous vapour in it. In England it rarely needs 30° of cooling, but in some hot climates it may need more than twice this amount.

The main essential characters of air, in as far as it is a gaseous body, have already been stated,† as well as the downward force it exercises, and the consequent compres-

---

* See *ante*, p. 96.   † See *ante*, pp. 78 and 147.

sion and greater density of its lower strata, owing to the pressure on them of the air above. The atmosphere is supposed to form a layer over the earth's surface of probably between forty and fifty miles in thickness. At the sea level, 100 cubic inches of air at 60° F. weigh about 30 grains, but at an elevation of 20,000 feet the pressure on them would be diminished one-half; for there is as much air in the lower $3\frac{1}{2}$ miles of the atmosphere as in all the superior portion. The weight of the atmosphere at any spot is tested by the barometer.*

As has been pointed out,† the sun's rays hardly at all raise the temperature of the atmosphere; it is warmed by the earth's surface, which is itself heated by the rays passing to it through the air. Then the superincumbent mass of air, becomes gradually warmed by convection,‡ and so currents upwards and downwards are produced. Now the earth's surface is very unequally heated, that of tropical lands being vastly hotter than at the arctic regions, and the air being most expanded by the warmer surface will rise to a greater elevation above it, than elsewhere.

But no fluid, either liquid or aëriform, can heap itself up. It must overflow, and then it will immediately be pressed upon by a rush of colder and therefore heavier air. Thus it is winds are produced, which are of the greatest utility in lessening extremes of temperature. There is a constant rise of warm air from the hottest regions of the earth, which then flows northwards and southwards at a high altitude towards the poles, while lower currents of cold air rush simultaneously towards the equator. This is, on a gigantic

---

* See *ante*, p. 82. † See *ante*, p. 98.
‡ See *ante*, p. 95.

scale, what takes place, on a minute scale, in every room where there is a fire, or which is in any way unequally heated. But the currents towards the poles after a time become so much cooler, while the equatorially rushing air becomes so much warmer, that they have generally been supposed to change places at about latitude 30°. Thence the air from the equator generally forms the lower current, till it approaches the pole, where it again rises amidst the irregularities of wind known as "polar gales" (Fig. 22).

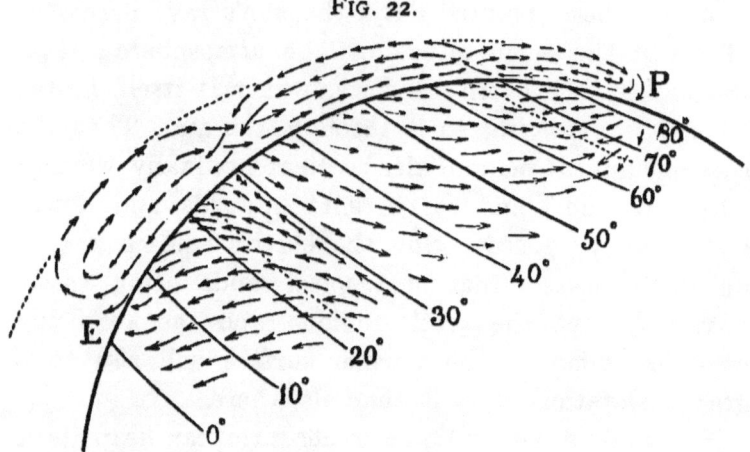

FIG. 22.

The action of different degrees of heat has great effect in determining the motion of the winds, but most conspicuous effects are also produced by another cause, as follows: The earth, as every schoolboy knows, revolves on its axis from west to east once in every twenty-four hours. Now, just as the body of every traveller in a carriage or a boat, participates in the velocity of the vehicle which carries him,* so the atmosphere more or less fully participates in the velocity of that part of the earth's surface it covers. But from the globular shape of

---

* See *ante*, p. 58.

the earth it is evident that while at the equator the surface of the earth has a rotatory motion of 1042 miles an hour, that of a circle very near one of the poles would travel at the rate of only ten miles an hour, and others still nearer at a less and less rate. Therefore polar currents (from the equator, towards the poles) must acquire an eastward direction, as their velocity east will grow more and more in excess of that of each tract of the earth's surface they pass over. On the other hand, the currents from the poles towards the equator will more and more fall below the velocity of each part of the rotating earth they successively come to. They will therefore lag behind it, and so appear to blow in a westerly direction and this constantly. These currents constitute the trade winds of the Northern and Southern Hemispheres. Near the equator they become neutralised, and the heated air ascends, and so we have an equatorial band of calms and occasional storms between the north and south trade winds. The trade winds are modified by the form of the continents they traverse, while they disappear north and south of latitude 30°. They disappear by giving place to the variable winds of the temperate regions with which we are familiar, and the conditions determining which are too complex to be further noticed here, the reader being referred for such knowledge to works on *Meteorology*.

The winds of the Indian Ocean known as "*monsoons*" are modifications of the trade winds due to the influence of vast masses of land. *Land and sea breezes* are due to the more equable temperature of the sea and the more ready heating and cooling of the surface of the land. During the day the tendency is for the air over the cooler sea to come in as a sea breeze, and replace the ascending current of air produced by the action of the sun's heat on the land. After sunset, on the contrary, the shores becoming cooler than the sea, we have an opposite effect, a *breeze from land to sea*. Hill and valley breezes also exist, partly due to mountain summits receiving more heat by day, and radiating it far more readily by night than do the lowlands, and partly to colder and therefore heavier air rolling down hill.

Rotatory storms, hurricanes, and typhoons may have a diameter of four or five hundred miles. Such movements are not so many transfers of the mass of air itself, but are rotatory movements of great velocity transmitted through its particles. They are supposed to be occasioned by a rapid movement of ascent communicated to air by some very heated spot of the earth's surface, such movement taking place through air either relatively at rest or moving in a contrary direction. This would be sufficient to occasion an incipient vortex which may be compared with the conical vertices before considered * as occurring in water. Rotatory storms wander about with a movement of translation which is slow when compared with the enormous rapidity of rotation, which may be more than ninety miles an hour. These aërial vortices proceed obliquely north and south from the region of the equator, turning from west to east, while their movement of translation, in the northern hemisphere, is westward till they have passed the region of the trade winds, then they turn eastwards.

There is an instrument for measuring the wind's force (*anemometer* or *anemoscope*) consisting of an upright U-shaped tube, containing a little water, with one end bent horizontally so as to face the wind. A scale for registering the height of the water is placed between the two limbs of the upright tube, the water being, of

---

* See *ante*, p. 74.

course, of the same height in each limb when there is no pressure. Then when the horizontal part of the tube is made to face the wind, its pressure will be registered by the amount of elevation thus produced, of the water, in the opposite limb of the tube. A gentle breeze will support 0.025 inches of water, a brisk gale, 0.5. A storm will sustain 3 inches, and a violent hurricane, 9 inches.

There are certain elevations and depressions which affect the whole aërial envelope of the globe, but these will be noticed in connection with "Ocean Tides."*

Water is commonly spoken of as being "fresh" or "salt" water. In fact, however, it always contains a greater or less quantity of foreign substances dissolved in it, but not, of course, chemically united with it. In the first place water—except water that has been boiled—contains a considerable quantity of air mixed up within it, and rain-water gathers in its descent, some of the air's soluble constituents, including carbonic acid and ammonia. The water of each river necessarily contains some of the salts of the springs which feed it, and it also contains the matters which it dissolves out from the material which it meets with in its course. Amongst the more noteworthy ingredients it thus acquires, are carbonate of lime and flint in a state of solution. Thus the Thames carries past Kingston daily not less than 1514 tons of solid substance (mainly derived from calcareous formations of Berkshire, Oxfordshire, and Gloucestershire), which includes more than 1000 tons of carbonate of lime. Sea-water notoriously contains a great deal of salt, with other chlorides and sulphates, and with some ammonia and iodine.

There are currents in the ocean due to differences of

---

* See *post*, p. 182.

temperature, as we have seen there are in the air. From the two extremely cold regions of the globe—the Arctic and Antarctic regions—cold ocean currents extend, in variously modified ways, towards the equator, while warm currents diverge north and south from the equatorial region towards the poles.

As every one is aware, there is a diurnal ebb and flow of the ocean which is known as "the tides," but their consideration will be deferred* till some words have been said about the celestial bodies, as otherwise they could not be understood.

The distribution of the earth's dry land has great effect on these currents and on the climates of the world generally. The greatly preponderating mass of dry land is situated north of the equator, and is divisible into two unequal sections—(1) one consisting of the continent of America, and the other (2) of the continents of Europe and Asia, with Africa. The latter section may be regarded as one great whole because, though Africa, but for the Isthmus of Suez, would be an island, the Mediterranean and Red Seas, which divide it from Europe and Asia, are very insignificant tracts of water compared with the great oceans, and each of these is bounded at its outlet by a very narrow strait. These two great sections descend southwards in three prolongations, dividing the earth's immense marine envelope of salt water into three oceans. The American, or New World, section, after becoming extremely narrowed towards the Isthmus of Panama, rapidly spreads out again into an enormous mass of land extending east and west, and then gradually narrows to Cape Horn, which reaches southwards to the 56° of south

---

* See *post*, p. 182.

latitude. Thus America divides the Atlantic from the Pacific Ocean. The old-world section of the earth sends southwards two prolongations. The first of these is formed by the African continent, which terminates at the Cape of Good Hope, and does not quite attain $34°$ of south latitude. The second prolongation is formed by Asia where it ends in the Malay Peninsula, but which does not quite reach the equator. It may, however, be said to be continued onwards by the mass of large and small islands which constitute the Indian Archipelago, and by the vast quasi-continent of Australia, which extends to a little beyond south latitude $39°$. The Atlantic ocean bounds Africa and Europe on the west, while the eastern shores of Asia and Australia are washed by the Pacific, the southern prolongations of Africa and Asia enclose between them the third, or Indian ocean, into the middle of the North of which Hindostan projects, running downwards till it ends at Cape Comorin in a little over $8°$ of north latitude.

With respect to the volume of the earth's marine liquid envelope compared with that of the land above its surface, it would seem that the former is more than forty times in excess of the latter. Also more than two thirds of the surface of the globe is covered by water, which, at the poles, assumes the form of ice. There is a much more extensive ice-cap on the South Pole than the North Pole, and no one knows how much land the south ice-cap may cover.

The globe can easily be mapped out in such a manner that one hemispherical map shall show very little besides water except the southern ice-cap. For if we make a map of one hemisphere, taking London as our centre, we shall have on it the maximum of land, while on the opposite hemisphere will be the maximum of water (Fig. 23).

The distribution of dry land is thus most irregular, as also is the degree to which the shape of any large portion of land is varied by deep indentations in, and far-reaching prolongations from, its shores.

Besides the British Isles and Iceland, those islands most desirable here to note (with a view to succeeding chapters) are Ceylon, Sumatra, Java, Borneo, Celebes, the Philippine and Japan Islands; the small islands Bali, Lambok, and Timor; Madagascar, with its distant outliers, Mauritius and Bourbon—all in the Indian Ocean; the Canaries, the Cape de Verde Islands, Ascen-

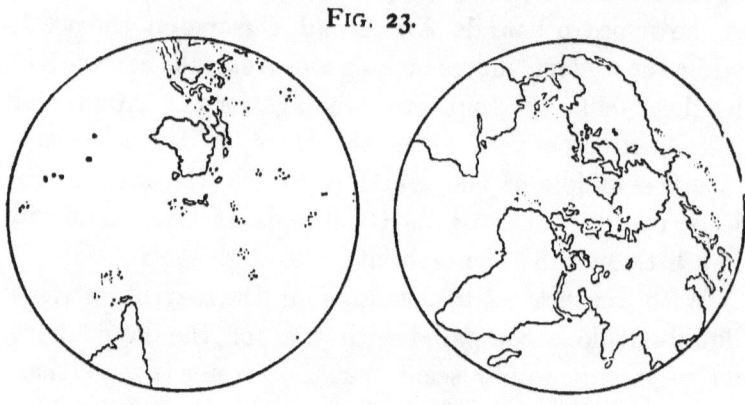

FIG. 23.

sion, and St. Helena; the West Indian Islands and Trinidad—all in the Atlantic Ocean; the Galapagos, Sandwich, and Society Islands, Fiji, New Caledonia, the Solomon Islands; New Zealand, New Guinea—all in the Pacific Ocean; and finally the Moluccas and Aru Islands, and Tasmania, which last, as it were, prolongs Australia to the south. Some of these are noteworthy as being separated by very deep water from land adjacent, while others are remarkable as rising in groups from a surface but little submerged, and separated by no great depth from an adjacent continent.

The greater oceans contain abysses which sink much deeper below the ocean's level than any mountains ascend above it. With these complex conditions of shore and sea-bottom, it is no wonder that ocean currents become thereby variously diverted from the courses they would take were they affected by nothing but variations of temperature. Nevertheless difference of temperature is the great cause of their existence, while one of their most noteworthy effects is the great change they can produce in land climates. Thus while cold currents sweeping down from the Greenland seas, carry ice and cold water southwards along the east coast of America, to 40° of north latitude, the equatorial current (proceeding westwards across the Atlantic and northwards to the Gulf of Mexico) and, its prolongation, the Gulf Stream (extending north-eastwards from the Mexican Gulf) carry warmth with their waters into western Europe and over the North Cape. Did a belt of land extend between Britain and Greenland, so as to intercept the passage of this warm stream (as the land bounding Behring Straits stays the passage northwards of the warm currents of the Pacific Ocean), we should then see the mountains of Scandinavia (like those in Greenland in nearly the same degree of latitude) permanently covered with ice and snow.

The elevations of the land—mountain chains—exert various effects on currents of air—the winds—analogous to those produced by the form of coasts on ocean currents. Mountains are sometimes very lofty, the highest in the world, those of the Andes, rise close on 30,000 feet, while Himalayan peaks exceed 27,000. The direction of mountain chains is also most influential. No line of mountains running east and west in North America, checks the descent of polar winds to the Gulf of Mexico. The great chain of the Andes runs from north to south very near the west coast of South America. Very different would be the effect of the trade winds did the Andes bound the eastern instead of the western coast of that continent.

These various modifications of aërial and ocean currents, cause the temperature of different places of the earth's surface and atmosphere and the degrees of their humidity (in other words, their climates) to vary independently of their degree of latitude, *i.e.*, their distance from the equator. Different, indeed, are the climates of mild Cornwall and frigid Newfoundland (though the latter is south of the former), and of Bordeaux and Halifax (in Nova Scotia)—both of nearly the same degree of latitude. Lines which connect places of similar temperatures are termed *isothermal lines*. They are much and irregularly curved, and therefore widely differ from the circles which mark degrees of latitude.

The evaporation of water over the earth's surface causes the atmosphere, as before pointed out,* to contain a greater or less quantity of aqueous vapour, sometimes even to become *saturated*, that is, to contain the greatest quantity it can possibly hold. The rapid condensation of this vapour produces rain.

When air laden with vapour blows from districts with a cooler climate to another which is warmer, it acquires a still greater capacity for aqueous vapour, and the result is that the clouds (which, as before said, are masses of minute particles of water), being dissipated into steam, vanish altogether.

On the other hand, when winds saturated with vapour pass into a cold region, torrents of rain may result.

---

* See *ante*, p. 148.

Such is the case when the trade winds, laden with vapour from the Atlantic, begin to ascend the slopes of the Andes, the summits of which—even at the equator—are clothed with perpetual snow at a height of sixteen thousand feet. At such an altitude, and at so low a temperature, the vapour assumes the solid form of crystalline water—*i.e.*, snow. The limit at which such snow can form itself on mountains—called the *snow-line*—descends as we recede on either side from the equator. In the Swiss Alps it is at about from 8500 to 8000 feet, but descends to 5000 feet on the Norwegian mountains. In the Arctic and Antarctic regions it gains, apart from the influence of winds, the sea-level. The mere cold of very great altitudes keeps lofty mountain-tops constantly below 32° F. In the equatorial regions, heat diminishes 1° for every 333 feet of ascent, and therefore above 15,700 feet it must freeze every night. But this lowering of the *snow-line* does not by any means take place with regularity, as the previously mentioned curvature of isothermal lines would alone be enough to show. Thus, Captain Cook found snow at the sea level in the island of South Georgia between 54° and 55° of south latitude. This is not more distant south from the equator than Durham is distant to the north of it. The piling up of snow by continual deposits on lofty mountains, causes the mass to force its way slowly down the highest valleys, through the action of gravity. Gradually solidifying, it forms *glaciers*, and they accommodate themselves to the various capacities of the depressions through which they travel, by breaking themselves up. But their broken fragments re-adhere with such speed that the mass often seems as if it squeezed through narrows without becoming fractured. On their road, they score and furrow the rocks they press against, and are powerful enough to scoop out large excavations, so deepening the depressions in the mountain sides and in the valleys which they traverse. The length of Swiss glaciers is sometimes twenty miles, their breadth occasionally two or three miles, and their depth 600 feet. They melt slowly as they descend, the water flowing in tunnels beneath them, and issuing from under ice arches at their lower extremity. Masses of rocks, or boulders, and many stones are carried along by them, accumulating at the glacier's lower termination, such accumulations being known as *moraines*.

In high latitudes, great masses of glacier will break off into the sea and float away to warmer climes, as *icebergs*, carrying with them large masses of rock and boulders with a large quantity of stones and mud. They have been seen so large as to be many miles in circumference and 300 feet high. Such a mass must be vast indeed, seeing that for every cubic foot above the sea's surface there must be eight cubic feet below it.

But the condensation which appears occasionally as snow, but generally as rain, takes place very unequally over the earth's surface. The tropics are most abundantly watered, parts of Brazil receiving annually as much as 270 inches of rain, and Cherra Poonjee, in Assam, 500 inches. On the east of the Andes, however, there is a narrow tract of land which is rainless because the constant western winds are drained of their water as they pass over the snow-capped mountains. That great desert, the Sahara, of Northern Africa, is rainless, because the moist winds and clouds it may receive from the Mediterranean find in it not a condenser but, on account of its heat, a vapouriser, which, as before said,* makes clouds vanish. The great tableland of Gobi, in

---

* See *ante*, p. 158.

Central Asia, is a rainless desert because the vaporous winds from the Indian Ocean are drained of their moisture ere they reach it over the Himalaya, while the mountains of China drain the humid winds which pass to it from the Pacific.

It is evident that districts far from the sea, and destitute of mountains, must possess less aqueous vapour in their atmosphere than others in the vicinity of the ocean, of seas, or of large lakes. This alone must make the climate of South America more humid than are the central regions of the Old World which are so much more distant from the shores of the ocean.

It is from rain that all the rivers of the world have their origin. For the springs from which so many arise are but the outcrop of the rain which has penetrated the soil till, having come to an impervious stratum, it can descend no longer, but must issue forth at the lowest level of the upper surface of such impervious stratum. In rain, all fresh-water lakes also have their origin, since they are produced by impediments in the rapid progress of rivers. There are, however, other lakes—such as the intensely salt Dead Sea—which have another origin, being remnants left behind by seas which have receded.

The largest rivers are in the New World; the *Amazon* runs a course of 3000 miles, and drains 1,500,000 square miles of country. The Mississippi of North America is almost if not quite as long, but drains a less extent of land. The Yangtse-kiang of China runs a course of 2500 miles; the Nile, the Ganges, and the Indus are also of great extent.

Water in the form of rain, rivers, streams and sea waves, is continually modifying the earth's surface, by destroying its more elevated parts. This modifying action is largely aided by ice, for, as we have seen,* water expands when it freezes, and thus it must enlarge any cracks and fissures into which it may have made its way and frozen. By these means the land is being continually torn down and carried off to be deposited either in estuaries, or at the mouths of rivers, or in the bed of the ocean. The mass of matter thus carried to the sea by some of the largest rivers is enormous. It has been calculated that the Ganges carries down every year as much land as could be carried down by 730,000 ships, each of 1400 tons burthen. The substances carried down by the Mississippi have formed at the mouth of that river, in the Gulf of Mexico, a deposit extending over an area of 30,000 square miles, and is known to be, at least in some parts, several hundred feet in thickness. A deposit thus formed at a river's mouth is generally more or less triangular in shape, on which account it is called a *delta*, from its resemblance in shape to the Greek letter so named. It is only when seas are more or less enclosed (as, *e.g.*, the Gulf of Mexico and the Mediterranean), or where the ocean currents are weak, that the transported materials are deposited so as to form deltas. Egypt largely consists of the delta formed by the Nile, and it has been calculated that not less than 17,000 years have been required for its formation. Deposits of this kind carried down by rivers into freshwater lakes, also form " deltas " therein.

The eroding action of water is notorious. When the gradient of a river is considerable (as is commonly the case in the upper courses of rivers), its excavating action is also considerable, and if such a gradient be maintained to the coast, the river will excavate a deep channel

---

* See *ante*, p. 91.

bordered by high land to its mouth—as in the Tyn and the Tweed. The excavation by a river of its own valley may leave here and there, high up in sheltered positions, accumulations of drifted materials, marking the level at which the river flowed at successive earlier periods.

Every child who has the opportunity of examining an undulating sandy surface after violent rains, may see clearly both the eroding and the depositing action of water. It needs but a much prolonged action of that kind to cause profound modifications of the earth's surface. There is no mountain which is not almost incessantly being in this way rendered more steep and precipitous, and thus the whole land of the globe constantly tends to be washed down by rivers, and spread out beneath the surface of the sea. But the finer débris of the land carried down incessantly into the sea by rivers, is, when the action of the river-water ceases, caught up by the great marine currents and swept to places more or less widely distant and out of the reach of tidal action.

The lowering of the earth's surface by the wear and tear of water is more or less counterbalanced by a slow, or rapid, upheaval of other parts of its surface through volcanic action. The number of burning mountains— active volcanoes—in the world may be estimated at about 300, and some of these give forth vast quantities of lava.* For example, in the Island of Hawai a burning deluge of lava broke forth in 1840 from the crater of Kilauea; it spread from one to four miles wide and reached the sea in three days, at a distance of thirty miles, and for fourteen days it plunged, in a vast fiery

---

* See *ante*, p. 147.

cataract a mile wide, over a precipice fifty feet high into the ocean.

Volcanoes are very unequally distributed over the earth's surface. One of the most considerable volcanic regions is that in the Andes (between Quito and Chili), and there is another in Mexico. A volcanic region of very great extent passes from the Philippines through the Moluccas, Timor, Lombok, and Bali into Java and Sumatra. Volcanoes have been known to resume their activity after a quiescence of centuries.

The slow upheaval and depression of different parts of the earth's surface have been proved by direct observations. The Andes have been rising century after century at the rate of several feet, while the region of the eastern Pampas has risen but a few inches. The land of Scandinavia, towards North Cape, rises above five feet in a century, and very many other instances could easily be adduced of slow secular elevation. Soundings often give good reason to suppose both that some rather distant lands once formed part of an adjacent continent, as also that islands, which by their proximity to some mainland might be supposed to have been previously united to it, have not really been so, but have only become nearer to it through a recent elevation of the mainland's coast. But however considerable such changes may here and there have been, it seems probable that the great oceans and continents have been permanent save for more or less considerable modifications of their margins and boundaries.

The science which treats of the structure of the earth, and the causes which have brought about the present condition of its surface, is the *science of geology*. The earth's crust, mainly composed of the mineral substances before noticed, is made up of superimposed masses of them, such superimposed masses being called *strata*. These strata consist of various, generally more or less horizontal, layers of different materials, and are generally composed of consolidated mud which has been deposited (in the way described) in fresh or salt water lakes, or in deep or shallow seas. But not all rocks are due to the agency of water. Many masses have been ejected in a molten state from volcanoes, and solidified either on the land's surface or beneath the sea, and therefore, in the latter case, under great pressure. Rocks which are thus due to volcanic agency are called *igneous* rocks. Those of them which have been formed under the sea are called *plutonic*; otherwise they are termed *volcanic*. Igneous rocks are not generally stratified, and they may be of all ages. Some, like those which form part of Snowdon and Cader Idris, are very old. Others, like those of Etna, and those which cover Herculaneum, are relatively quite recent. Deposits may have undergone four kinds of change : (1) they may have undergone a mere process of drying (as with sands); or (2) drying accompanied by pressure (as with sandstone); or (3) with chemical action in addition (as with highly crystalline rocks like that called gneiss*); or (4) a change may have been produced by infiltration. Thus rocks may be infiltrated by iron, lime, or silica, producing ferruginous, calcareous, and silicious sandstones and conglomerates, which last are sometimes called "pudding stones," and consist of fragments of rock cemented together.

The strata thus forming the crust of the earth are supposed to be from sixteen to eighteen miles thick; but no boring has yet extended to even one mile in depth, and indeed has scarcely exceeded 3000 feet. The total

---

* See *ante*, p. 147.

depth, therefore, is purely a matter of inference from the arrangement, superposition, and inclination of the different strata, as seen at or near the surface.

After penetrating a moderate distance, the temperature of the earth's interior has been found to augment, the greater the depth explored, at the rate of $1°$ Fahrenheit, sometimes for every 45 feet and sometimes for every 70 feet approximately.

The various strata of which the earth's crust is composed were, of course, deposited at successive times, and the time of the deposition of each is called its "period" or "epoch." But for subsequent disturbance, the most ancient strata would always be deepest, and superposition would, in all cases, plainly indicate relative novelty. As it is, we have often to examine carefully in order to discover the real order of deposition, but this once discovered, depth is equivalent to age, and *vice versâ*. The uppermost and most recent accumulations of sands, clays and gravels, form what are called the "recent deposits," and these are not counted as forming any part of the proper geological strata, and are not represented in ordinary geological maps, but are there disregarded. The strata beneath these recent deposits are classed in three great groups, belonging respectively to three great epochs. The deepest and most ancient group comprises the strata called *Primary* or *Palæozoic*. The second or middle group of strata is called *Secondary* or *Mesozoic*. The uppermost and least ancient group consists of strata called *Tertiary* or *Cainozoic*. The "recent deposits" really belong to this last-mentioned group, and we may be said to be still living in the Tertiary period, which has succeeded the only two earlier periods of which as yet we have evidence—namely, the secondary and the primary periods or epochs.

THE NON-LIVING WORLD

FIG. 24.

Periods	Systems	Formations	
Quaternary	RECENT	Terrestrial, Alluvial, Estuarine, and Marine Beds of Historic, Iron, Bronze, and Neolithic Ages	
	PLEISTOCENE (250 ft.)	Peat, Alluvium, Loess, Valley Gravels, Drickearths, Cave-deposits, Raised Beaches, Palæolithic Age, Boulder Clay and Gravels	
CAINOZOIC / Tertiary	PLIOCENE (100 ft.)	Norfolk Forest-bed Series, Norwich and Red Crags, Coralline Crag (Diestian)	
	MIOCENE (125 ft.)	Œningen Beds Freshwater, &c.	
	EOCENE (2,000 ft.)	Fluvio-marine Series (Oligocene), Bagshot Beds, London Tertiaries } (Nummulitic Beds)	
SECONDARY OR MESOZOIC	CRETACEOUS (7,000 ft.)	Maestricht Beds, Chalk, Upper Greensand, Gault	
	NEOCOMIAN	Lower Greensand, Wealden	
	JURASSIC (3,000 ft.)	Purbeck Beds, Portland Beds, Kimmeridge Clay (Solenhofen Beds), Corallian Beds, Oxford Clay, Great Oolite Series, Inferior Oolite Series, Lias	
	TRIASSIC (3,000 ft.)	Rhætic Beds, Keuper, Muschelkalk, Bunter	
PRIMARY OR PALEOZOIC	PERMIAN or DYAS (500 to 2,000 ft.)	Red Sandstone, Marl, Magnesian Limestone, &c. } Zechstein, Red Sandstone and Conglomerate, Rothliegende	
	CARBONIFEROUS (12,000 ft.)	Coal Measures and Millstone Grit, Carboniferous Limestone Series	
	DEVONIAN & OLD RED SANDSTONE (5,000 to 10,000 ft.)	Upper Old Red Sandstone, Devonian, Lower Old Red Sandstone	
	SILURIAN (3,000 to 5,000 ft.)	Ludlow Series, Wenlock Series, Llandovery Series, May Hill Series	
	ORDOVICIAN (5,000 to 8,000 ft.)	Bala and Caradoc Series, Llandeilo Series, Llanvirn Series, Arenig and Skiddaw Series	
	CAMBRIAN (20,000 to 30,000 ft.)	Tremadoc Slates, Lingula Flags, Menevian Series, Harlech and Longmynd Series	
	EOZOIC— ARCHÆAN (30,000 ft.)	Pebidian, Arvonian, and Dimetian, Huronian and Laurentian	

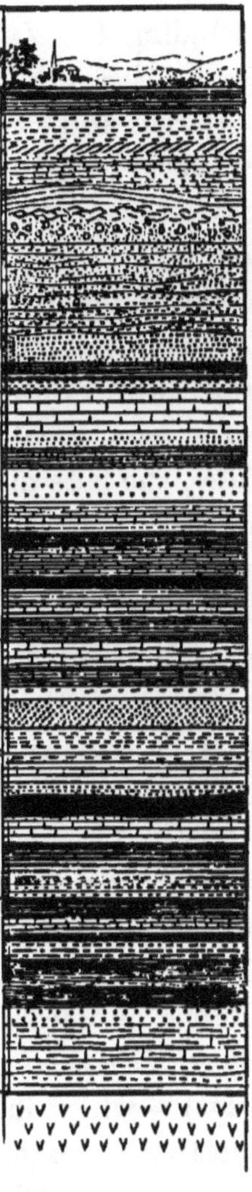

(After Dr. Woodward, F.R.S.)

Each of these three great groups of rocks is made up of a certain number of subordinate groups of strata or *formations* grouped in systems. Thus the Palæozoic, or Primary rocks, are made up of the *Eozoic-archæan* (including the *Laurentian* formation), *Cambrian, Ordovician, Silurian, Devonian, Old Red Sandstone, Carboniferous* and *Permian* systems. The Laurentian rocks are very largely developed in Canada, and are some 30,000 feet in thickness. The Cambrian rocks are from 15,000 to 20,000 feet thick, and are well seen in the Longmynds of Shropshire, and near Bangor, Harlech, and St. Davids in Wales. The Silurian strata (sandstones, clays, limestones, and igneous rocks) are of very great thickness, and form a large part of Wales, the lake district of England, Southern Scotland, and some parts of Ireland. The Devonian system is exemplified in Devon and Cornwall, and the Old Red Sandstone rocks of Ireland, Scotland, and Wales. The Carboniferous system includes the *Carboniferous limestone* and the *Coal measures*. The latter consists of seams of coal and layers of sandstone and slate; such alternations indicating oscillation of level. The Permian system is of moderate thickness, and mainly consists of *magnesian limestone* associated with many slates and beds of conglomerate. In England it is chiefly found skirting the coal-fields from Durham to Derbyshire. The *Mesozoic* or Secondary rocks, are made up of *Triassic, Jurassic, Neocomian,* and *Cretaceous* systems; the first (*Trias*) —which includes strata known as the "New Red Sandstone"—extends in England from Devon to Yorkshire, and is largely developed in Cheshire. The Jurassic rocks contain all formations from the *Lias* and the *Oolite*, to the *Purbeck beds*. The Lias extends from Lyme Regis, obliquely north-east to Whitby. The Oolite also extends between the north-east and south-west of England. To the upper portion of the Jurassic rocks belong the *Solenhofen slates* of Bavaria. The Cretaceous and Neocomian systems include the *Wealden*, the lower and upper *Gault*, and the *Chalk*. The Wealden is well seen in the *Greensand*, the south-east of England, where (in Kent, Surrey, Sussex, and the Isle of Wight) it is considered to represent the delta of a large ancient river. The Chalk formation ranges from Lyme Regis to Flamborough Head, and also forms both our North and South Downs. It and the Maestricht beds terminate the series of *Mesozoic* formations, and a great break exists between it and the Tertiary formations which follow. The break between them seems to be partially bridged over in North-Western America by certain beds known as the "Lignite Series." The *Cainozoic*, or Tertiary rocks consist of three systems — the *Eocene*, the *Miocene*, and the *Pleiocene*. Eocene rocks underlie both Paris and London, and form very important deposits in North America. The Miocene formation is widely distributed in Europe and the North American Continent, but is very slightly represented in Britain. The igneous rocks which form the Giant's Causeway, the Island of Staffa, of Mull and others, belong, however, to the group. The Pleiocene formation is extensively distributed in Europe, Asia, and the United States. In England it is represented by the Norfolk and Suffolk "Crag." The later, *Pleistocene*, rocks — the so-called *Quaternary* strata — include the deposits found in ancient caves in Europe and those thrown down during what is known as the *Glacial Epoch*. That such a period of intense cold prevailed over Northern and Central Europe and the greater part of North America, in geologically recent times, is shown by the evidence of prodigious glaciers,* which appear to have scooped out valleys, and scored the surface of hill and dale in those regions. Blocks of stone or "boulders" are often found scattered about there, and seem to have been transported by ice, sometimes for very great distances.

The various strata which thus form the crust of the earth contain, in different degrees of rarity or abundance, certain objects which are known as "fossils." Amongst the mass of materials carried down by rivers and deposited along their course or in their deltas, or at the bottom of the sea, are numerous relics of that kind. When such a relic becomes entombed, it often happens that particle by particle of its substance is replaced, particle by particle, by mineral matter (ferruginous, calcareous, or silicious) till we have a complete representation—technically called a "*pseudomorph*"—of the original in such new material.

There are five kinds of fossils :—

(1) Objects preserved unchanged, or little changed so that they retain the greater part of their own mineral material matter; (2) substitutes or "pseudomorphs"; (3) moulds, *i.e.*, deposits which present the impressions made by objects, all other evidences of which have disappeared. Thus impressions of hailstones made in a soft surface of mud, have often been preserved by subsequent delicate layers of deposit. Then, the whole having become hardened, the shape of such impressions when laid bare by the geologist, may show us to-day the direction in which the wind blew when those hailstones fell at some unimaginably distant period of past time; (4) casts of moulds, *i.e.*, solid matter which has taken the place of whatever bodies may have produced the moulds and then

---

* See *ante*, p. 159.

disappeared—such would be the solid matter filling up the hollows formed by the hailstones and so replacing them; (5) casts of hollow structures, *i.e.*, mineral matter which has filled cavities in the interior of fossils and so formed internal casts (as "moulds" are external casts) of fossils. For further information about the earth's crust the reader is referred to special works on geology.

Having now considered the effects of heat, motion, and chemical changes upon the globe as a whole, it remains to speak of the influence exercised upon it by electricity and magnetism, as well as of the external sources of its light and heat, and the effects produced upon it by them through gravity.

We have seen * that the motion of either electricity or magnetism, circulating round an axis, develops the other force *along* that axis. Now there is a constant flow of electric currents around the earth from east to west, and, besides this, the unequal heating of the surface of the globe in each twenty-four hours has its necessary electrical consequences.†

The result of the circulation of electricity round the globe is to make the world itself a huge magnet with two opposite magnetic poles. These poles are not far from the poles of the earth's rotation—the north and south poles. The northern magnetic pole is near Hudson's Bay, and the opposite one is amidst the Antarctic ice; but their positions slowly change and revolve round the earth from east to west.

It is the presence of these poles which has enabled navigation to be aided by the magnetised needle of "the mariner's compass." This magnetic needle is so placed on a pivot that it can turn freely in any direction, and

---

* See *ante*, p. 127. † See *ante*, p. 124.

it constantly points towards the magnetic north pole—or rather its long axis always coincides with the line joining the magnetic poles—unless it be made to deviate by some local magnet, or mass of iron in its vicinity. But as the magnetic poles do not coincide with the earth's poles of rotation, and as the imaginary circles or "meridians" (real as drawn on maps) which mark degrees of longitude, all pass through the poles of rotation (the geographical north and south poles), the long axis of the magnet will not coincide with such "meridians" along any great circle. Theoretically, there ought to be one such circle of coincidence, but, owing to the fact that in the globe (as in other natural magnets) magnetism is, more or less, irregularly distributed, the line of coincidence is also irregular. This circle extends from near Hudson's Bay, through the United States, Cuba, Jamaica, and Brazil, to the South Magnetic Pole, whence it is continued on through Australia, China, and Siberia to the North Magnetic Pole, near Hudson's Bay, whence we started. The further the magnetic needle may diverge from this line, the greater, of course, will be its divergence from a geographical meridian, and such divergence is called its *variation*. The magnetic needle, while constantly pointing towards its pole, undergoes slight changes, daily, monthly, and yearly, while every now and then it undergoes sudden and irregular disturbances, indicating what are called *magnetic storms*, and these seem to extend over the whole earth.

The magnetic needles may be exactly balanced horizontally, and yet able to lower one end, or, as it is technically termed, *to dip*. The nearer such a needle is brought towards either magnetic pole, the more it will dip, and at either such pole it will be vertical, while along a line equidistant from both, it will be horizontal. The "dip" of the needle is subject, like its variation, to daily, monthly, and yearly changes, as well as to sudden storms.

The external sources of the world's heat and light are, of course, the "heavenly bodies"—the sun, moon, and stars, the study of which constitutes the *science of astronomy*.

The earth, as every one now knows, is not only a sphere, revolving on its axis daily, but also accomplishes, together with its satellite, the moon, an annual revolution round the sun. It is therefore a planet, *i.e.*, one of those various other spheres which also revolve round the sun, together with their satellites; and which planets, with certain comets, and clouds of more or less relatively minute bodies, called meteoroids, constitute a planetary, or solar, system. This system again is but one of many systems of suns (it may be with or without attendant planets) which make up, together with some dark globes and masses of gases or vapours—termed *nebulæ*—the visible sidereal universe.

The various bodies of this universe, which vary immensely as to size, are continually changing their relative positions according to the mechanical laws of dynamics, already noted, and the force of gravity. These bodies are all material bodies, and we have seen * that all such are (by gravity acting between them) drawn together directly as their masses and inversely as the squares of their distances. It is the study of the relative movements of the heavenly bodies which has revealed to us the universality, so far as we have been able to test it, of the law of gravitation.

* See *ante*, p. 66.

The grand result of this energy is that the planets of our system, and doubtless those of other systems, revolve round suns or central bodies, in ellipses, variously attended by satellites, which in turn revolve around their respective planets. In some distant systems there may be more than one sun. Thus for our purpose we may consider the universe as divided into two parts: (1) The sun with its attendant bodies, *i.e.*, the *solar system;* and (2) The rest of the universe which can by any means be made visible to us—nebulæ, and all the bodies called "fixed stars" because they have for us no obvious movement. We class the latter, for convenient description, in groups termed *constellations*, such as that of the Great Bear (familiarly known as Charles's Wain) and others. Such groups, however, have no natural connection but are only associated together on account of their conspicuousness and apparent proximity.

Our own solar system is rushing at the rate of ten thousand miles every half-hour in the direction of the constellation, known as *Lyra*, and no doubt all the other suns or fixed stars are similarly in motion, although their great distances make such movements inappreciable. The known universe, or cosmos, is made up of bodies variously composed of solid, liquid, or gaseous matter, and these bodies differ greatly in density, some, as before mentioned, being but masses of vapour, the "nebulæ."

The cosmical bodies shine either by self-emitted light (as does our sun and the variously distant stars) or by reflected light, as do the planets and satellites of our solar system and, probably, multitudes of planets of other systems, though some planetary bodies themselves may be very faintly self-luminous.

All our heat is derived from the sun, and also almost

all our light, since the moon and planets reflect their light on us, the light of the distant self-luminous stars being quite insignificant in comparison with the direct and reflected solar light. The surface and atmosphere of the sun form a region of intense energy and heat. Just as aqueous vapours ascend from the surface of the earth and, after condensation, fall back on it as showers of rain, so in the sun metallic vapours are continually ascending, to be condensed and fall down in showers of red-hot metal, amidst flames of hydrogen,* thousands of miles high.

Now, as we have seen,† light travels at the enormous speed of 186,330 miles in a second, through whatever intervenes between, and connects together, all the planetary and stellar bodies. The hypothetical substance, ether,‡ before noted as the medium of light,§ must, if indispensable for luminous energy, be universally diffused wherever light is transmitted. Wherever light can travel, there must then be ether; and light can travel through every known interval, and through the most perfect vacuum which we can make. Therefore, neither can we make a real vacuum, nor can there be any between us and the most distant visible star. The distance from us of the nearest visible fixed star (*Alpha centauri*) being about 272,000 times greater than that of the sun—which itself is about 92,700,000 miles away—the time which light must take in passing from it (the nearest fixed star) to our eyes is three years. Even the light of the sun takes eight minutes to come to us, and that of the most distant known planet, Neptune, takes five hours. The light of the most distant visible stars probably takes centuries to reach us, so that in what we

---

* See *ante*, p. 139. † See *ante*, p. 103.
‡ See *ante*, p. 93. § See *ante*, p. 102.

see of them we only see what existed more than one hundred years ago.

This great and uniform speed of light combined with the motion of the earth, causes what is known as the *aberration of light*.* It is common to all the heavenly bodies, and causes them all to appear a little out of their true place. While the light is travelling from any star towards an observer, the observer himself is being simultaneously carried along with great rapidity by the earth.

The result is that if he directs his telescope exactly towards a star, the light which enters his instrument will strike the side of it before it can reach his eye. This has therefore to be allowed for, and the instrument accordingly directed a little in advance of the object observed, just as a sportsman has to shoot in front of a running hare, to allow for its change of place during the passage of a shot.

The distances and sizes of the nearer heavenly bodies, and the size of the earth itself, have been ascertained by the application of geometrical principles, and mainly that one which teaches us that two triangles, however different in size, are in other respects exactly similar to each other provided the angles of one are the same as the angles of the other.

Thus if we require to know the exact distance of some object on that side of a wide river which may be opposite to us, we may ascertain it as follows: We must select two spots A and B on our side, and measure the length of the straight line between them. Then planting ourselves at A we must observe, with an instrument made for that purpose, a distant object, O, and ascertain the angle formed by the line AB with that which

---

* Before referred to, p. 103.

passes from our eye at A to O. Next we must plant ourselves at B, and similarly observe the angle formed by AB with a line from B to O. Thus we may easily obtain a triangle which will enable us to tell the distance of O. To do this we must draw on paper a straight line *ab*, and from its two ends draw two other lines such that the angle formed at *b* by one of them shall be the same as the angle we observed at the point B, while the angle formed at *a* shall be equal to that we observed at the station A. Let these lines be *ax*

Fig. 25.

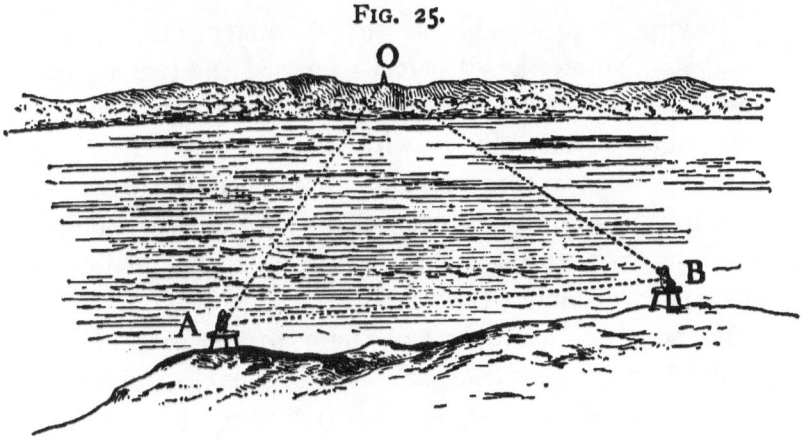

and *ay* respectively; then if we prolong them enough they will meet at some point *p*. Then the small triangle on paper, *apb*, will be similar to (*i.e.*, equiangular with) the large triangle formed by the straight line AB and the two lines respectively extending from A and B and meeting at the distant object O. Therefore the length of AB must bear the same proportion to *ap* as the line AB bears to AO and by measuring the actual lengths of the three lines of the triangle *apb* and that of the line AB, it is easy by the simple arithmetical process known as "the rule of three," to ascertain the dimensions of the two lines AO and BO and so the distance of the object O.

By an extension of this principle to the observation of the angles formed by lines passing from any one spot on the earth's surface to some fixed star (a process which cannot be explained here), it has been ascertained that the earth is a globe with a circumference of 24,840 miles. We can ascertain this because the distance of the fixed stars is, as we have seen, so enormous that they remain for our observations always practically in one relative place.

Having ascertained the earth's dimensions we can treat its diameter as we before supposed the two ends of the measured line AB to be treated; and so ascertain the angles formed by it with lines passing from its extremities to the moon, and when once the distance of a body is known, we can readily find its *size* and *vice versâ*, by the simplest application of the before-mentioned principle of "similar triangles."

By careful observations of the apparent path of the planet Venus across the sun's disc (in what is known as the transit of Venus), as seen from two spots on the earth's surface, it was determined in 1761 and 1769 that the sun's distance from the earth was from 93,274,000 to 96,432,000 miles. This being ascertained, it became easy to ascertain the size of the earth's orbit, which then provided an enormously larger and more useful base for triangular measurements. Our measured line AB might now be taken as 190,000,000 miles (such being the diameter of the earth's orbit), observations from the opposite extremities of which, as to the angles formed by it with more distant heavenly bodies, served for further investigation of dimensions and distances. But so distant are the fixed stars that even with this enormous base, the amount of their remoteness remains unascertainable save in some thirty instances, as (with these exceptions) they present no appreciable difference of position, or, as it is called, no *parallax*, when viewed from opposite points of the earth's orbit.

The sun is 852,900 miles in diameter, and 1,252,700 times the volume of the earth.

Our satellite, the moon—which is only 238,813 miles distant from us, and has but a diameter of 2,160 miles—circles round us in about four weeks and turns once on her own axis during each such revolution. Therefore the same side of the moon is ever turned towards the earth. The so-called changes in the moon, of course, simply result from the different amounts of her surface which are, at different times, illuminated by the sun's rays. Similar changes are also shown by the planet Venus.

The moon appears to be devoid of both air and water, or if such substances exist there, they seem to have retreated into the moon's interior, and give no signs of their presence on its much scarred surface.

As to the shape of the paths followed by the earth and the other planets in their revolution round the sun, they follow precisely the same laws as regulate any body so moving round another that it cannot fall to the surface of the latter. We have already seen * that any body projected from a point external to the earth's surface, and with a certain velocity, would be constrained to revolve round it in an ellipse and that its *radius vector* must always pass through equal areas in equal times and always in the same plane. This is the precise law of the earth's, and of all the other planets', revolutions round the sun.

---

* See *ante*, p. 65.

Every planet, moreover, not only describes an elliptical orbit, but one whereof the sun's centre is one of the foci. These laws were discovered by Kepler, who also found out that the time occupied by a planet in revolving round its orbit is proportional to the square root of the cube of the mean (*i.e.*, average) diameter of its orbit. The times (in days) which the different planets take so to revolve are : Mercury, 88 ; Venus, 225 ; the Earth, 365 ; Mars, 687 ; Jupiter, 4333 ; Saturn, 10,759 ; Uranus, 30,687 ; Neptune, 60,181.

As to the relative size of the sun and planets, if the Earth be represented by a pea, Venus will also be so represented ; Mercury by a grain of mustard seed ; Mars by a large pin's head, Jupiter by an orange, Uranus by a cherry, while the Sun would need a sphere 4 feet in diameter to represent it. As to degrees of density, Mercury is about twice as dense as the Earth, which itself is about five and a half times as dense as water. Jupiter, on the other hand, is but a quarter of the density of our globe.

As might be expected, though some of the planets differ greatly from the earth in density and other physical conditions, there is a substantial resemblance between them which is sometimes carried very far. Thus our nearest neighbour—after the moon—the planet Mars, appears to be so like our earth as to have not only its tracts of land and water but also caps of polar ice.

The planets of the solar system, with their satellites, move round the sun in the same direction, with the exception of the satellites of Uranus and the solitary attendant on Neptune. These move in a retrograde direction, and the former are also very exceptional in that their orbits are nearly perpendicular to the plane of the earth's orbit.

The planets, like the earth, revolve on their axes while they go round the sun, and their satellites revolve round the planets they attend more slowly than such planets revolve on their own axes. An exception, however, occurs in the case of the planet Mars, one of the satellites of which circulates round it in less than a third of the time that planet takes to revolve on its own axis.

Since the earth's orbit is elliptical, our globe must be nearer the sun at one time (in one part of its path) than at another. It is furthest from the sun during our summer and nearest in winter, but its axis slants in such a manner that its north pole inclines towards the sun in summer and away from it in winter. Therefore the earth's northern hemisphere receives the sun's rays most directly when it is furthest from it, and least so when it is nearest to it. This is why our summer season is the warmest, while the same period is the coldest for the globe's southern hemisphere. In spring and summer the condition is intermediate, hence the four " seasons."

During each spring and autumn, our obliquely inclined globe is for a short time in such a position that the illuminated half of its surface is situated symmetrically as regards the poles. This is the period of the *equinoxes*, or of equal night and day all over the globe.

As the northern pole becomes more and more inclined towards the sun, its daylight is prolonged till (*e.g.* North of North Cape) the sun never sets, while in winter it never rises. The opposite condition of course obtains at the Antarctic, or south pole.

The circuit described by the earth in its path round the sun is constantly changing to a small degree. It alternately approximates to, and diverges further from, a truly circular path. It is evident that when its degree of eccentricity (*i.e.*, its greatest departure from a circular path) is greatest, the pole which when thus most distant from the sun, is also inclined away from it, must then endure a very exceptional degree of cold. Of course, at such a time, the opposite pole and hemisphere must be very exceptionally warm.

The direction also of the earth's axis slightly varies, each pole describing a circle (comparable with that described by the summit of a revolving teetotum) in a very long period of time; that is to say, in nearly twenty-six thousand years. This movement is spoken of as the *precession of the equinoxes,* because each change in the position of the earth's axis necessarily changes the position, in the earth's orbit, of the spot where equal day and night are experienced all over the globe.

Evidently from the great proximity of the moon to the earth, there must be an energetic action of gravity between them, and this energy must produce its most conspicuous effect upon what is at once most easily moved and can most plainly be seen to be so moved.

That which is thus most easily and evidently moved, is the earth's liquid investment—the ocean. The effect of the action of gravity between the earth and the moon, is made manifest by the tides—the moon raising up the surface of the sea as it revolves round the earth. This action is, of course, modified by that of the sun, but on account of the enormous distance of the latter, its action is much less than that of our satellite. Did the moon act alone, and were the earth perfectly spherical and everywhere covered with a uniform depth of water, the moon would so attract the water that there would be one great wave, or heap of water, directly beneath it, and another on the opposite side of the globe, the water in the interspace being correspondingly depressed.

Thus it is (1) when the sun and moon are exactly opposite each other, the world being between them (which is the time of full moon), and (2) when they are on the same side, or in conjunction, the world being opposite both (which is the time of new moon), that their combined actions produce the very high tides. The intermediate periods give rise to the relatively slight, or "neap," tides.

The ellipticity of the earth's orbit and also that of the moon, cause variations in the proximity of these bodies to each other and to the sun. It is when the sun and moon are thus nearest to us, that we get the very high or "spring" tides.

The various configurations of tracts of land and water, and their geographical positions, variously modify the times and degrees of elevation in the tides of different places. Thus, for example, the Mediterranean is an almost tideless sea.

An action, which is so visible with respect to the ocean, must also take place in the atmosphere, and thus, as before mentioned,* there must be atmospheric tides, though they do not make their existence conspicuous.

The phenomena known as "falling stars" are due to the attraction to the earth of minute cosmical bodies, *meteorites*, which it encounters in its path round the sun. These bodies afford us the plainest proof that the same chemical substances exist in the solar system, external to this earth, as exist in the earth itself. But the careful study of the spectrum—*spectrum analysis*—which can be obtained from the light of the stars, tends to show us that a similar identity of materials exists between the substances which compose our own planet and

---

* See *ante*, p. 153.

those which enter into the composition of even the most distinct stellar bodies. Thus the action of gravity and the energies known as light, heat, motion and chemical action, as also, doubtless, those activities termed electric and magnetic, seem to be diffused throughout the visible universe. The same is doubtless the case with other energies and influences, if such there be, which remain, as yet, undiscovered and unknown.

These few elementary notions with respect to the heavenly bodies, and the effects of their energies upon the globe we inhabit, are all which space permits us to give. For more than an introduction to such first elements of science, the student must be referred to works devoted to the exposition of the science of astronomy.

CHAPTER VI

THE LIVING WORLD

WE have now, in our study of elementary science, to make a great step in advance. The objects which have next to occupy our attention, like those which have previously occupied it, conform, as a matter of course, to the laws of number and of mechanics, and serve as vehicles for the physical forces herein before noticed. But the objects we have now to consider also possess additional powers—powers of which the whole non-living world is entirely destitute. Every one knows that plants grow and multiply, and that animals not only grow and multiply, but have their feelings also. Our dog can plainly hear and see us, and has his sensation of pleasure and of pain, as well as his emotions of hope and joy, of fear and grief. But so highly organised an animal cannot serve to set forth the subject of this chapter. For by "the living world," the whole mass of creatures, from the humblest green thread of conferva, or most microscopic fungus, to the gigantic gum-tree and the far-spreading banyan; and from the hardly perceptible animalcule to the humming-bird, the condor, the tiger, the whale, the monkey and man.

To study so enormous a mass of forms with any completeness is beyond the power of any single human being.

It is therefore absolutely impossible in this chapter to

do more than indicate what the various branches of such a study are, and to briefly portray some of the most elementary and indisputable facts which concern living beings. For everything beyond this, students are referred to the many special treatises which exist on each of the numerous sub-divisions of the study of living creatures considered as one whole.

The study of that living whole—the science which includes the study of all living things—is termed *biology*, while *botany* treats only of plants, and *zoology* exclusively of animals.

The living creatures with which we are familiar, have various active powers, while every animal or plant has a certain structure of its own. The most casual observation suffices to show that a fowl, a lobster and an oyster, a rose-tree, a Scotch fir, a mushroom and a sea-weed, have each of them a structure more or less different from that of each of the others. There is a science which deals with obvious structural differences, namely *anatomy*. That word, taken by itself, generally refers to the study of the structure of the human body, while the structure of other animals, compared therewith, is spoken of as *comparative anatomy*. But plants have also their anatomy, though their structures are much simpler than that of most animals.

The material frame of an animal or plant may soon be seen to consist of different kinds of substances. Thus a cat's body* will be found to consist of fur, skin, flesh, nerves, bones, &c. Similarly, a tree, such as an elder, will be seen to be made up of woody substance, solid and

---

* Readers are referred to the author's work on the cat (John Murray) as a complete introduction to all branches of the study of living things.

hollow fibres, pith, leaf-substance, &c. Now these various substances are called *tissues*, and each, when examined by the microscope, is found to be composed of very small structures, which are for the most part known as "cells." The study of tissues and their minute structure is called *histology*.

Anatomy shows us that living creatures are composed of various parts of organs, each made up of certain tissues. Thus the cat, like man, is provided with a mouth to receive food, teeth to divide and crush it, a stomach to digest it, an alimentary canal and liver, a heart and blood-vessels, a brain and nerves, &c., each being composed of the various tissues which enter into its composition. An oak also has its stem, containing many tubular vessels, roots which spread into the soil, and leaves which expand and expose themselves to the sun's rays. All these different parts or organs concur in sustaining the life of the creature which possesses them, and therefore the life of its various other organs.

In this way each organ is reciprocally "end" and "means"; on which account living creatures are commonly spoken of as *organisms*. Organs are also united together into groups which are called "*systems.*" Thus the mouth, stomach, intestine, liver, &c., constitute what is called the *alimentary system;* the heart and vessels form the *vascular* or *circulating system*, and the brain and nerves the *nervous system*.

Now all these various component parts of living creatures have their respective activities or *functions*, which minister to, and are subsumed in, the life of the creature itself. The study of these various functions or activities is termed *physiology*.

Each "cell" has its own activity, as has each "tissue" (formed of cells of one kind) and each "organ" (formed of different tissues) and each "system" (consisting of various organs), the harmony of all the "systems" resulting in the life of the creature whereof such "systems" are component parts.

Certain lowly organisms consist but of a single cell, and are therefore spoken of as *unicellular*.

We know that an animal (*e.g.*, the cat) eats, digests, and so nourishes itself, circulates its blood, breathes, forms (*i.e.*, *secretes*) saliva, bile, &c.; feels, moves to and fro, and may become the parent of another generation. These various life activities, or functions, are respectively known as *alimentation, nutrition, circulation, respiration, secretion, sensation, locomotion,* and *generation*. Every one knows also that plants and animals grow, as also that plants generally spring from seeds, and birds from eggs.

The study of that particular growth which takes place in a plant from the seed, and in the bird from the formation of the egg, is called *embryology*.

But animals and plants have very definite relations with space and time. Monkeys and armadillos do not exist in a wild state in England, and kangaroos are found nowhere, naturally, save in the Australian region; nor are cacti, which are so common in Mexico, found wild in Scotland. There is, then, a science of the *geography of organisms*. They have also definite relations to past time. A multitude of animals and plants which existed in Eocene* times do not live now, and this is still more the case with the reptiles whose remains are found fossil in the secondary strata, while it is clear that many animals which now live did not do so in those earlier periods. Thus, organisms have

---

* See *ante*, p. 169.

definite relations with past time as well as with space; and it is evident that as age has succeeded age, there has been a process of replacement in vegetable and animal forms, new kinds having come into being one after another. It seems also evident that in the earliest ages the world was entirely devoid of living creatures.

Furthermore, animals and plants have also definite relations with each other. If a beast of prey finds its way into a region peopled with creatures good for food, it will increase and multiply to their detriment; while the most peaceful animal will suffer, by the introduction, into a limited area, of creatures which are rivals because they feed upon the same food, the supply of which will, sooner or later, be insufficient for all. Here the reader may ask, if the world was once without life, whence did life come?—and what is life? Also, since new kinds have replaced older ones which disappeared, the question naturally arises, how did new kinds arise?

But in a work like the present one (which is but an introduction to the elements of science), the consideration of such questions would be out of place. They would be as much out of place as would be a consideration of the questions "What is heat?" and "What is light?" As to the latter questions we have provisionally noted certain useful working hypotheses. Similarly, since up to the present day, life has not been evolved by us from inorganic matter, we may, as a working hypothesis, adopt the belief that life is the energy of a peculiar form of force which exists differently in each different kind of organism, and that this force is a main agent in the development of new kinds. As to the first introduction of life on the surface of this planet our reason is as yet entirely in the dark.

The number of all the various kinds of living

creatures is so enormous that it would be impossible to study them profitably, were they not classified in an orderly manner. Therefore the whole mass has been divided, in the first place, into two supreme groups, fancifully termed *kingdoms*—the "animal kingdom" and the "vegetal kingdom." Each of these is subdivided into an orderly series of subordinate groups, successively contained one within the other, and named *sub-kingdoms, classes, orders, families, genera* and *species.* The lowest group but one is the "genus," which contains one or more different kinds termed "species," as *e.g.*, the species "wood anemone" and the species "blue titmouse." The lowest group of all—a species—may be said to consist of individuals which differ from each other only by trifling characters, such as characters due to difference of sex, while their peculiar organisation is faithfully reproduced by generation as a whole, though small individual differences exist in all cases.

The vegetal, or vegetable, kingdom, consists of the great mass of flowering plants, many of which, however, have such inconspicuous flowers that they are mistakenly regarded as flowerless, as is often the case with the grasses, the pines, and the yews. Another mass, or sub-kingdom, of plants consists of the really flowerless plants, such as the ferns, horsetails (Fig. 26), lycopods, and mosses. Sea and fresh-water weeds (*algæ*), and mushrooms, or "moulds," of all kinds (*fungi*), amongst which are the now famous "*bacteria*," constitute a third and lowest set of plants.

The animal kingdom consists, first, of a sub-kingdom of animals which possess a *spinal column*, or backbone, and which are known as *vertebrate* animals. Such are all beasts, birds, reptiles, and fishes. There are also a variety of remotely allied marine organisms known as *tunicates*, sea-squirts, or *ascidians* (Fig. 27). There is, further, an immense group of *arthropods*, consisting of all insects, crab-like creatures, hundred-legs and their allies, with spiders, scorpions, tics and mites. We

Fig. 26.

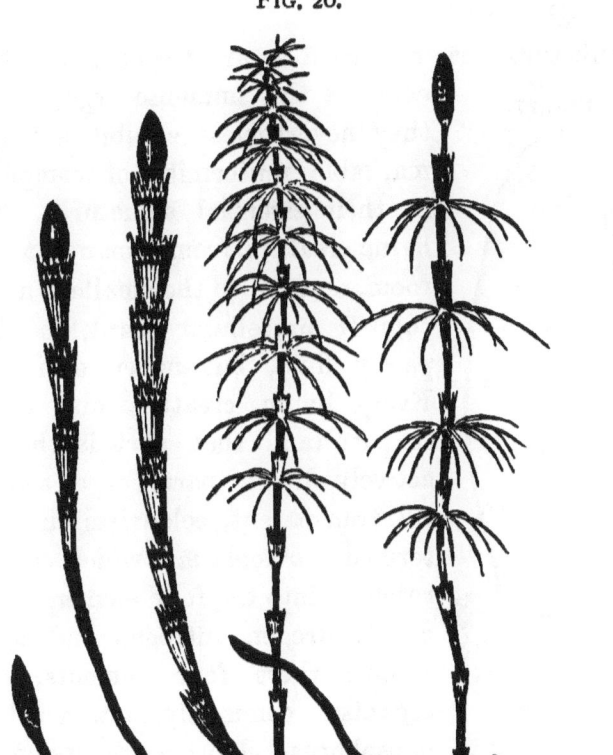

HORSE-TAIL (*Equisetum drummondii*).

have also the sub-kingdom of shell-fish or *molluscs*, including cuttle-fishes, snails, whelks, limpets, the oyster, and a multitude of allied forms. A multitudinous sub-kingdom of worms also exists, as well as another of star-fishes and their congeners. There is yet another of zoophytes, or polyps, and another of sponges, and, finally, we have a sub-kingdom of minute creatures, or *animalculæ*, of very varied forms, which may make up the sub-kingdom of *Protozoa*, consisting of animals which are mostly unicellular.

Multitudinous and varied as are the creatures which compose this immense organic world, they nevertheless exhibit a very remarkable uniformity of composition in their essential structure. Every living creature, from a man to a mushroom, or even to the smallest animalcule or unicellular plant, is always partly fluid, but never entirely so. Every living creature also consists in part (and that part is the most actively living part) of a soft, viscid, transparent, colourless substance, termed *protoplasm*, which can be resolved into the four elements,* oxygen, hydrogen, nitrogen and carbon. Besides these four elements, living organisms commonly contain sulphur, phosphorus, chlorine, potassium, sodium, calcium, magnesium and iron.

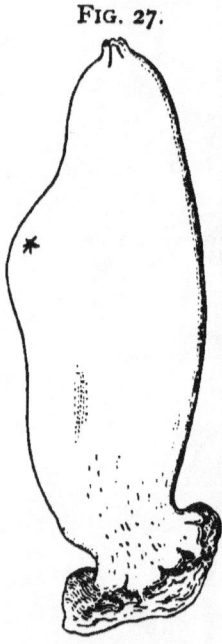

FIG. 27.

A TUNICATE
(*Ascidia*).

In the fact that living creatures always consist of the four elements, oxygen, hydrogen, nitrogen and carbon, we have a fundamental character, whereby the organic and inorganic (or non-living) worlds are to be distinguished. For, as we have seen, inorganic bodies, instead of being thus uniformly constituted, may consist

---

* See *ante*, pp. 138-140.

of the most diverse elements and sometimes of but two or even of only one.

Again, many minerals, such as crystals,* are bounded by plain surfaces and, with very few exceptions,† none are bounded by curved lines and surfaces, while living organisms are bounded by such lines and surfaces.

Yet again, if a crystal be cut through, its internal structure will be seen to be similar throughout. But if the body of any living creature be divided, it will, at the very least, be seen to consist of a variety of minute distinct particles, called "granules," variously distributed throughout its interior.

All organisms consist either (as do the simplest, mostly microscopic, plants and animals) of a single minute mass of protoplasm, or of a few, or of many, or of an enormous aggregation of such before-mentioned particles, each of which is one of those bodies named a "cell" (Fig. 28, p. 195). Cells may, or may not, be enclosed in an investing coat or "cell-wall." Each cell generally contains within it a denser, normally spheroidal, body known as the *nucleus*.

Now protoplasm is a very unstable substance (as we have seen many substances are whereof nitrogen ‡ is a component part), and it possesses active properties which are not present in the non-living, or inorganic world. In the latter, differences of temperature will produce motion in the shape of "currents," as we have seen § with respect to masses of air and water. But in a portion of protoplasm, an internal circulation of currents in definite lines will establish itself from other causes.

Inorganic bodies, as we have seen, will expand with

---

* See *ante*, p. 143.
† Spathic and hematite iron and dolomite are such exceptions.
‡ See *ante*, p. 139. § See *ante* pp. 150–157.

heat, as they may also do from imbibing moisture; but living protoplasm has an apparently spontaneous power of contraction and expansion under certain external conditions which do not occasion such movements in inorganic matter.

Under favouring conditions, protoplasm has a power of performing chemical changes, which result in producing heat far more gently and continuously than it is produced by the combustion of inorganic bodies. Thus it is that the heat is produced which makes its presence evident to us in what we call "warm-blooded animals," the most warm-blooded of all being birds.

Protoplasm has also the wonderful power of transforming certain adjacent substances into material like itself—into its own substance—and so, in a sense, creating a new material. Thus it is that organisms have the power to nourish themselves and grow. An animal would vainly swallow the most nourishing food if the ultimate, protoplasmic particles of its body had not this power of "transforming" suitable substances brought near them in ways to be hereinafter noticed.

Without that, no organism could ever "grow." The growth of organisms is utterly different from the increase in size of inorganic bodies. Crystals, as we have seen,* grow merely by external increment; but organisms grow by an increment which takes place in the very innermost substance of the tissues which compose their bodies and the innermost substance of the cells which compose such tissues; this peculiar form of growth is termed *intussusception*.

Protoplasm, after thus augmenting its mass, has a further power of spontaneous division, whereby the mass

---

* See *ante*, pp. 143 and 144.

of the entire organism whereof such protoplasm forms a part, is augmented and so growth is brought about.

The small particles of protoplasm which constitute "cells" are far indeed from being structureless. Besides the nucleus already mentioned there is a delicate network of threads of a substance called *chromatin* within it, and another network permeating the fluid of the cell substance which invests the nucleus, often with further

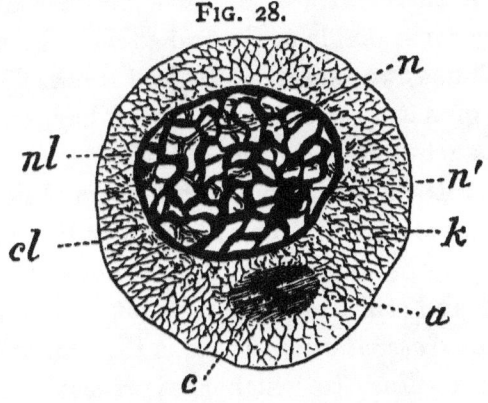

FIG. 28.

CELL FROM A SALAMANDER.

n, nucleus; *n'*, nucleolus embedded in the network of chromatin threads; *k*, network of the cell external to the nucleus; *a*, attraction-sphere or archoplasm containing minute bodies called centrosomes; *cl*, membrane enclosing the cell externally; *nl*, membrane surrounding the nucleus; *c*, centrosomes.

[*Drawn by Mr. J. E. S. Moore.*]

complications. These networks generally perform (or undergo) a most complex series of changes every time a cell spontaneously divides. In certain cases, however, it appears that the nucleus divides into two in a more simple fashion, the rest of the cell contents subsequently dividing—each half enclosing one part of the previously divided nucleus. It is by a continued process of cell division that the complex structures of the most complex organisms is brought about.

The division of a cell, or particle of protoplasm, is indeed a necessary consequence of its complete nutrition.

For new material can only be absorbed by its surface. But as the cell grows, the proportion borne by its surface to its mass, continually decreases; therefore this surface must soon be too small to take in nourishment enough, and the particle, or cell, must therefore either die or divide. By dividing, its parts can continue the nutritive process till their surface, in turn, becomes insufficient, when they must divide again and so on. Thus the term "feeding" has two senses. "To feed a horse," ordinarily means to give it a certain quantity of hay, oats or what not; and such indeed is truly one kind of feeding. But obviously, if the nourishment so taken could not get from the stomach and intestine into the ultimate particles and cells of the horse's body, the horse could not be nourished and still less could it grow. It is this latter process, called *assimilation*, which is the real and essential process of feeding, to which the process ordinarily so called is but introductory.

Protoplasm has also the power of forming and ejecting from its own substance, other substances which it has made but which are of a different nature to its own. This function, as before said, is termed *secretion;* and we know the liver secretes bile and that the cow's udder secretes milk.

Here again we have an external and an internal process. The milk is drawn forth from a receptacle, the udder, into which it finds its way, and so, in a superficial sense, it may be called an organ of secretion. Nevertheless the true internal secretion takes place in the innermost substance of the cells or particles of protoplasm, of the milk-gland, which particles really form that liquid.

But every living creature consists at first entirely of

a particle of protoplasm. Therefore every other kind of substance which may be found in every kind of plant or animal, must have been formed through it, and be, in fact, a secretion from protoplasm. Such is the rosy cheek of an apple, or of a maiden, the luscious juice of the peach, the produce of the castor-oil plant, the baleen that lines the whale's enormous jaws, as well as that softest product, the fur of the chinchilla. Indeed every particle of protaplasm requires, in order that it may live, a continuous process of exchange. It needs to be continuously first built up by food, and then broken down by discharging what is no longer needful for its healthy existence. Thus the life of every organism is a life of almost incessant change, not only in its being as a whole, but in that of all its protoplasmic particles also.

Prominent among such processes is that of an interchange of gases between the living being and its environment. This process consists in an absorption of oxygen and a giving-out of carbonic acid, which exchange is termed *respiration*.

Lastly, protoplasm has a power of motion when appropriately acted on. It will then contract or expand its shape by alternate protrusions and retractions of parts of its substance. These movements are termed *amœbiform*, because they quite resemble the movements of a small animalcule which is named *amœba* (Fig. 29).

Such is the ultimate structure, and such are the fundamental activities (or functions) of living organisms (so far as they can here be described), from the lowest animalcule and unicellular plant, up to the most complex organisms and the body of man himself.

It has been explained how it is that organisms of complex structure become such by means of a spontaneous division and multiplication of their component cells. Many unicellular organisms also divide themselves into two equal halves, which each grow as large as was the previously undivided cell. Thus new individuals are generated in the simplest fashion imaginable.

Other forms send forth more or less delicate prolongations of their substance, at the ends of which minute cells, termed *spores*, are produced, each of which, under favourable circumstances, will grow up into the form of the parent which produced it.

Minute water-weeds, which may consist of but a

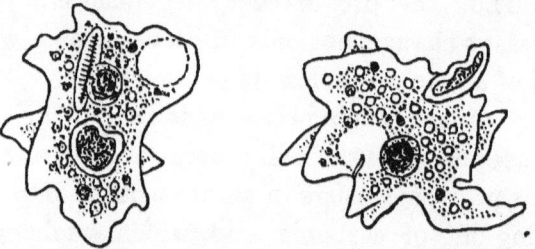

AMŒBA SHOWN IN TWO OF THE MANY IRREGULAR SHAPES IT ASSUMES. (*After Howes.*)

The clear space within it is a contractite vesicle. The dark body is the nucleus. In the right-hand figure there is shown a particle of food, passing through the external surface.

thread-like single series of cells (*confervæ*) will, when two such threads are adjacent, produce spores by *conjugation*. For this purpose, processes from opposite cells of two such thread-like plants, will grow forth, meet, and then blend together their protoplasmic contents. The result of this process is the production of a spore, which will afterwards grow into another conferva—thread. Those multicellular fungi known as ."puff-balls," give forth, when ripe, such a multitude of minute "spores" as to resemble a puff of smoke—

whence their name. That part of a mushroom, which rises from the ground, is also a "spore" producer.

The cells of plants are enclosed in a cell-wall which consists of a substance known as *cellulose*, and which contains no nitrogen.

The unicellular water-weeds (*algæ*) contain a green substance termed *chlorophyll*, which somehow enables them, during daylight, to dissolve carbonic acid and retain its carbon, while they let its oxygen go free.

FIG. 30.

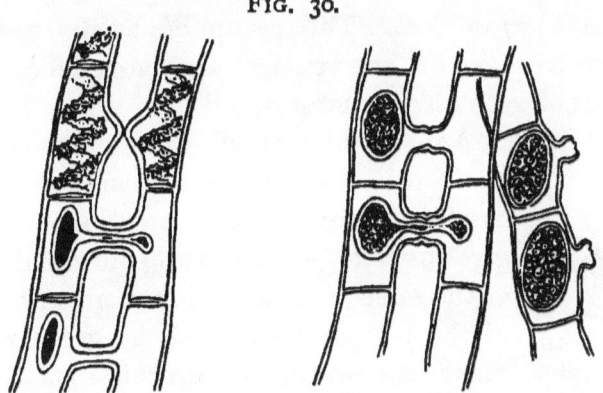

PORTIONS OF FIVE THREADS OF *Confervæ*, GREATLY MAGNIFIED.

Showing the cells of which they are composed and also the protrusion and blending of prominences of some of the cells, and the passage of the protoplasmic cell-contents through two of the protrusions which have blended.

Such plants can, in this way, nourish themselves and live on inorganic substances. Fungi, on the other hand, which possess no *chlorophyll*, and are not green, cannot do this, and therefore require for food living matter, or matter which has lived. Both kinds of plants respire, and therefore both, in breathing, take oxygen from the air and give out into it carbonic acid from their own substance in exchange, but it is only the green plants which give out oxygen, when feeding, by absorbing the carbon of the atmospheric carbonic acid.

All the higher plants resemble, in this respect, the green unicellular ones, and therefore enormous volumes of oxygen are given forth by the vast forests and extensive grassy plains which clothe the earth's surface, as also by the masses of seaweed in the ocean.

There is a sea-weed called *Lessonia* which forms submarine forests, with stems like the trunks of ordinary trees, while the sea-weed called *Macrocystis* may attain a length of 700 feet. The group of small sea-plants known as *Floridiæ* are amongst the most delicate and elegant of vegetable structures.

FIG. 31.

PROTOCOCCUS, with two Vibratile Cilia.

Animals, like fungi, cannot dissolve carbonic acid, absorb its carbon and let its oxygen go free; but in breathing they all give forth large quantities of carbonic acid, at the same time absorbing a great amount of the oxygen given forth by green plants. This is the reason why animals soon become suffocated when enclosed in anything which denies them a supply of fresh air. When so enclosed, they soon exhaust the supply of oxygen available, and die of suffocation, because they can then no longer exchange their carbonic acid for it.

Some of the lower plants move about in water with much activity, as for example, does the alga called *Protococcus* (Fig. 31), which moves by means of two minute hair-like processes termed *vibratile cilia*, which effect repeated lashings, the cause of which is, as yet, quite unexplained. Another very simple plant is called *Volvox* (Fig. 32), and consists of a spheroidal aggregation of cells bearing outwardly a multitude of projecting cilia, the regular lashings of which cause the spheroidal whole to rotate, while young spheroidal aggregations are formed within the parent.

Some plants, as the well-known sensitive plant, move their leaflets on being touched. Others will move parts of their flowers, either for the purpose of setting seed, or for protection, or for some other reason (as the pimpernel will close its flowers under a clouded sky), or to project their seeds to a considerable distance, as is the case with certain balsams. There are two kinds of plant, however, which are quite exceptional in the movements they will effect. The first of these is the sundew of the genus *Drosera*. The upper surfaces of its leaves bear certain hair-like processes which can discharge a tenacious fluid. Any insect, settling upon the leaf, is apt to be caught by these processes, which bend over it and bathe it in the fluid they distil. The other plant is called Venus's fly-trap (Fig. 33) (*Dionæa*). Its leaves terminate in two rounded expansions, joined by a median hinge, and they bear strong bristles along their margins. When an insect alights on this structure, the two rounded expansions snap sharply together and imprison it.

FIG. 32.

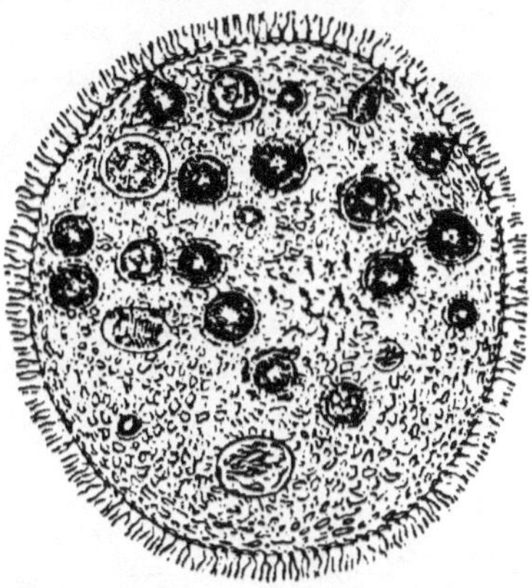

Volvox globator, MUCH MAGNIFIED. Showing *ribratile cilia* on its surface and numerous young contained within it.

Such phenomena are, indeed, different from any to be met with in the non-living world, though they are, of course, nothing to the complex movements of animals. Everybody knows that, as a rule, plants hardly move at all, while active locomotion is the common characteristic of the animal kingdom.

It is absolutely impossible in this work to give even the merest outline of all the principal groups of plants and animals. For further information the reader is referred to works devoted to botany and zoology. We must content ourselves with selecting a flowerless and a flowering plant as examples of vegetable life, adding thereto a brief notice of a few leading types from the animal kingdom.

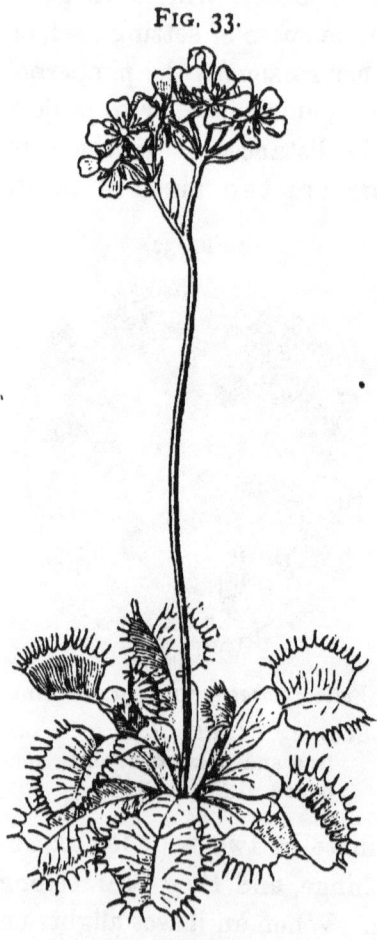

FIG. 33.

VENUS'S FLY-TRAP (*Dionæa muscipula*).

Showing flower above and the fly-catching leaves round its base.

Passing by, therefore, the lichens, liverworts, mosses, lycopods, and horsetails, and leaving the student to seek a knowledge of them elsewhere, we will briefly consider the common bracken fern, called by botanists *Pteris aquilina* — *Pteris* being the name of the genus to which it belongs, and *aquilina* indicating which species it is of that genus.

It is an organism of considerable size made up of a multitude of cells variously transformed to constitute the different "tissues" whereof the entire plant is made up.

The fern consists of an axial portion corresponding to the stem of most plants, which runs along underground, giving off at intervals the parts which appear above ground and are called *fronds*. These fronds correspond

Fig. 34.

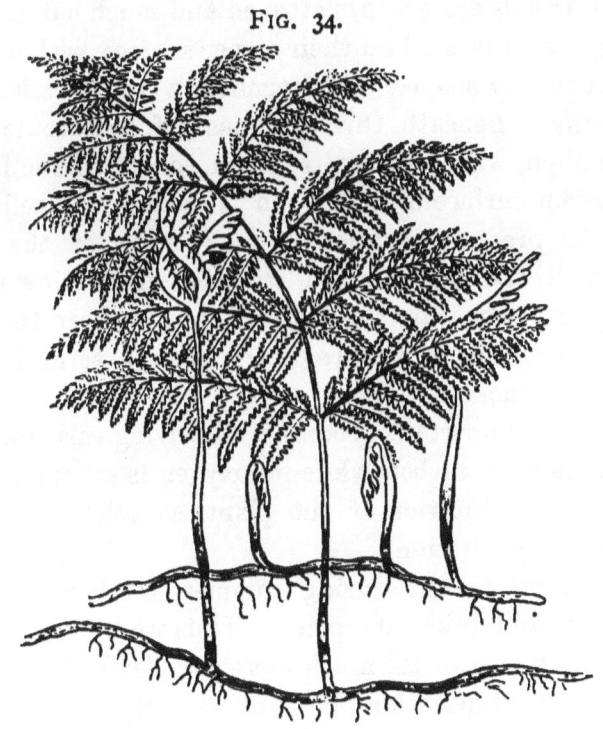

BRACKEN-FERN (*Pteris aquilina*).
Showing fronds springing from underground stem, which gives out rootlets beneath.

with the leaves of ordinary plants; and all leaves, however modified, are distinguished, by the term *foliar organs*, from stems, which, however modified, are called *axial organs*. Roots, in the form of filamentary processes, are given off from the under surface of the creeping subterranean stem or *rhizome*. The latter is formed of differently shaped cells aggregated in masses so as to appear as bands of different colours when it is cut transversely. These bands consist of fibres and tubes amidst large polygonal cells containing many granules of starch—which is a non-nitrogenous substance that plants produce abundantly.

The fronds are green, flattened and much sub-divided expansions, invested on their upper surface with a layer of irregularly shaped cells forming what is called the *epidermis*. Beneath this is a mass of cells containing chlorophyll, which gives its green colour to the frond. The under surface of the frond is coated with cells and hair-like processes, while between many of the cells are small openings, termed *stomata*, which allow air to enter and penetrate the cavities (left between the cells which form the substance of the frond), termed *intercellular passages*.

Thus the important process of dissolving carbonic acid and fixing its carbon while its oxygen is set free, takes place in the interior of the plant, as well as does the process of respiration.

If a frond be cut during summer, another will soon grow up, and takes its place. Thus we have a process of reparative growth, much more complex and complete than is the reparation of a mutilated crystal, or plant-like aggregations of crystals, which is the nearest approximation to true "growth" that is to be met with in the non-living world. In the fern, however, we have a mode of growth to which nothing in the non-living world makes even the faintest approximation.

Under the margin of a full-grown frond will be found a groove containing a series of small brown bodies, each of which is called a *sporangium*, because it is a little membranous bag that contains *spores*, which bag bursts and scatters the spores when ripe. Each such spore is a double-walled cell containing protoplasm. If it falls on a suitable surface it gives forth a prolongation, the cells

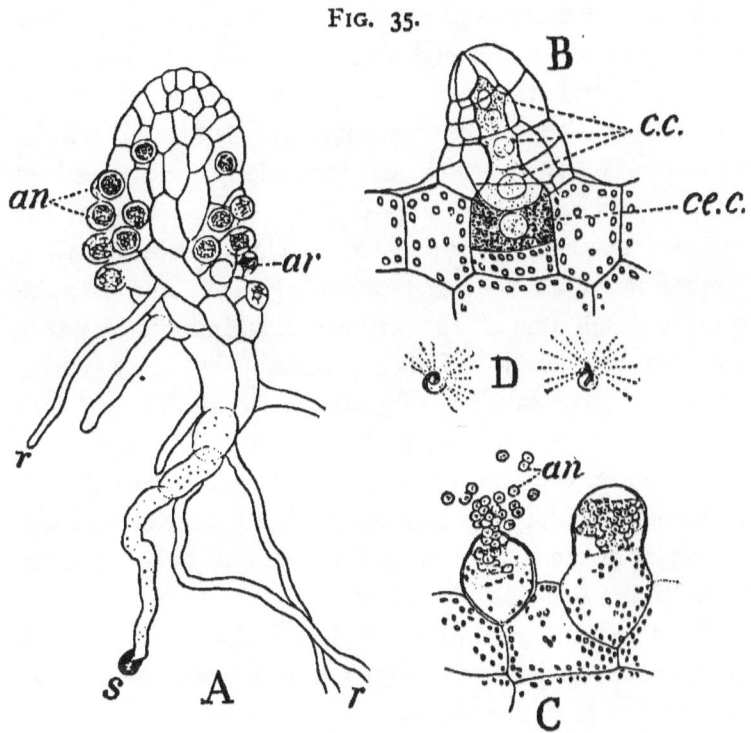

FIG. 35.

PROTHALLUS AND ITS PARTS MAGNIFIED.

A Whole prothallus.
   s Spore.
   r Root filaments or rhizoids.
   an Antheridia.
   ar Archegonia.
B An archegonium greatly magnified.
   cc. Canal cells.
   ce.c. Central cells.
C Two antheridia greatly magnified.
   an Antherozoids escaping from one of them.
D Antherozoids still more magnified.

formed from which divide and subdivide till they form a small cake-like expansion termed a *prothallus*, which sends down root-fibres from its under surface.

Then there appear on the prothallus a few rounded

prominences, each of which is called an *antheridium*, and contains within it cells, from each of which there comes forth a little spiral body bearing many cilia, by means of which it can move actively about. This little corkscrew-like structure is termed an *antherozoid*. Meantime on the prothallus other prominences have appeared, each of which is called an *archegonium*, and is a small cellular tube (the walls of which are formed by "canal cells"), open at its apex, and exposing to view a central cell at the bottom of the tube. This distinct cell is denominated an *oosphere*, or *embryo* cell, and remains quiescent till one of the antherozoids finds its way to it and blends with it. It is a process which may remind us of the process of conjugation between different cells of confervæ and is termed *impregnation*. When it has taken place, the embryo cell divides and subdivides till it forms an incipient rhizome with its rootlet and this young stem soon throws up a frond, and so the original form of the fern is reproduced.

Thus we have not only growth and change, but a cycle of changes, and what is sometimes spoken of as "an alternation of generations." Thus we have:

(1) A fern-organism, which produces

(2) A prothallus-organism, from which

(3) A fern-organism again results. As before said, nothing even faintly approximating to this cycle of changes, occurs anywhere amongst bodies devoid of life.

Let us next examine the structure and life-processes of a bean plant (*Vicia faba*). Here we have an axial organ, or stem, which is not a rhizome but grows upwards from the soil, giving forth roots from its base. From opposite sides of the stem spring forth foliar organs, in the form of green leaves, and also branches, or ramifications of the axis, which again bear green foliage

THE LIVING WORLD 207

FIG. 36.

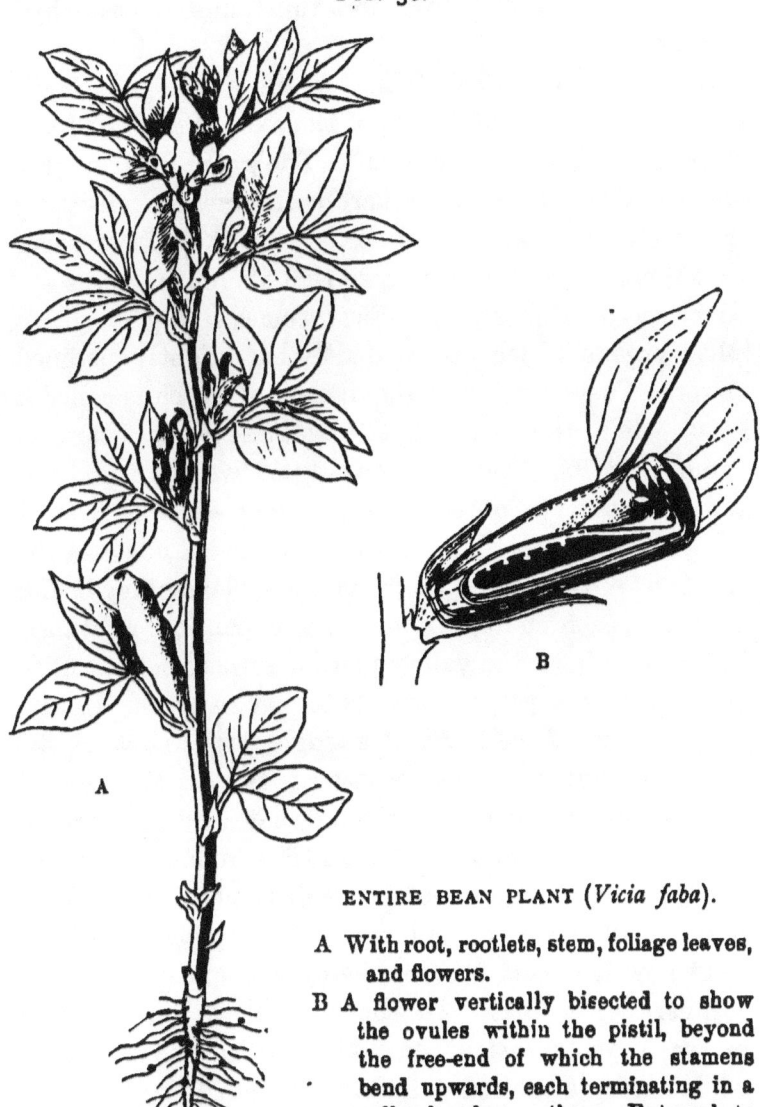

ENTIRE BEAN PLANT (*Vicia faba*).

A With root, rootlets, stem, foliage leaves, and flowers.
B A flower vertically bisected to show the ovules within the pistil, beyond the free-end of which the stamens bend upwards, each terminating in a pollen-bearing anther. External to the stamens two of the petals are to be seen, external to the base of which are two sepals of the calyx.

leaves on either side. These foliar leaves are essentially like the frond in structure and function save that they bear no spores.

But there are other foliar organs which grow forth around delicate ramifications of the axis, and which form what we know as "the flower." The green foliar organs are separated from each other by interspaces consisting of successive portions of the axis; but these interspaces are suppressed in the flower, so that its foliar organs are all closely approximated. First comes a ring of green foliar organs which are evidently but slightly modified leaves. There are five of them, each of which is called a *sepal*, while the whole five constitute what botanists name the *calyx*. Then comes a ring of five more modified and differently coloured leaves, each of which is called a *petal*, the whole five petals forming what is known as the *corolla*. Within the corolla are ten filamentary bodies (*stamens*), each ending in an oval expansion, or *anther*, which contains a fine yellow powder termed *pollen*. Each particle of this powder is called a *pollen-grain*. Lastly, in the centre of the flower is a single body known as the *pistil*, whereof the upper portion is termed the *style*, at the extremity of which is a somewhat modified surface, spoken of as the *stigma*. The pistil is hollow and, if cut open, will be found to contain small bodies named *ovules*, which are attached to its inner surface by short stalks.

The ovule is very far from being a simple cell, like the oosphere of a fern. In the first place it is enclosed in two coats, except at one point, called the *micropyle*, while amongst the cells, of which the small body consists, is one of larger size, termed the *embryo sac*, within which again two thickenings arise, from the upper one of which the future plant develops itself subsequent to impregnation, while the lower one affords it nourishment.

The agent of impregnation is a pollen grain, every one of which consists of a fluid particle of protoplasm, enclosed in a membrane or wall, which is almost always double.

As the yellow pollen dust flies about, one grain sooner or later is wafted to the stigma of the pistil, or, as in very many plants, is carried there by insects in search of food, who have previously dusted themselves with pollen in visiting other plants. As soon as the pollen-grain finds itself there, the particle of protoplasm, enclosed in its inner coat, passes through its outer one and descends through the intercellular spaces of the style till it reaches an ovule. Then it passes through the micropyle, or minute aperture left in the ovule's two coats, and penetrates its substance till it reaches the embryo sac which it impregnates. Then the upper thickening within it (before spoken of) grows and develops, from a mere formless mass of cells, into a miniature plant with a minute rudimentary stem and root and two pairs of leaves. Two of these leaves are termed the *cotyledons*. They become of relatively enormous size, enclosing the minute plant between them and constituting the great mass of the seed we familiarly know as " a

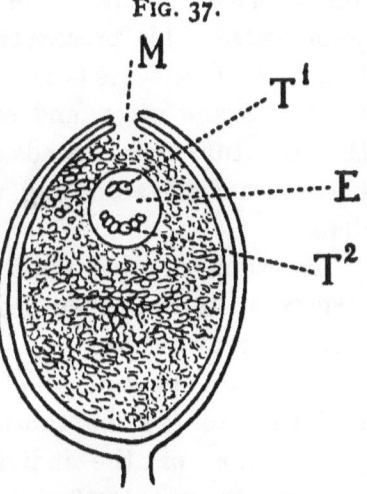

FIG. 37.

DIAGRAM OF AN OVULE.
Vertically bisected and showing the two coats which enclose it.

E Embryo sac.
M Micropyle or minute opening through which the pollen-tube enters.
T^1 Upper thickening of cell-contents, in which the embryo of the future plant takes its origin.
T^2 Lower thickening.

bean." Meanwhile the pistil, which encloses the growing ovules, itself rapidly enlarges into a *pod*, which, when ripe, bursts and sets free the ovules which have now become *seeds*, after which it decays.

When the seed, the bean, finds its way to the earth, under fixed conditions of warmth and moisture, it *germinates*. This process of germination consists in the bursting of the seed-coat by the swelling cotyledons, which become green and emerge as fleshy leaves, while the miniature stem ascends and the little root descends, both meantime absorbing nutriment from the cotyledons.

The stem of the adult bean plant is of complex structure, its component cells having become modified into different tissues.

As to the physiology of the bean plant, it may be divided into (1) the functions which minister to the preservation of the individual and (2) those which concern the preservation of the race.

Liquid is absorbed by the roots and part of it evaporates, by what is called *transpiration*, from the surfaces of the leaves, whence air and vapour are conveyed inwards—through their stomata.* Thus a sort of circulation also takes place. Liquid ascends between the fibres of the stem, being drawn up through evaporation from the leaves and being pushed up by the absorption of the roots. Water with suitable salts † in solution (nitrates of potassium or of calcium and sulphates of iron or of magnesium) is absorbed by the roots, while carbon is fixed by the decomposition of carbonic acid, during daylight, by the green leaves. These processes result in the formation of cellulose,

---

* See *ante*, p. 204. † See *ante*, p. 134.

starch, sugar and other products. The ascending fluids permeate from cell to cell upwards, while the products of carbon-fixation permeate from cell to cell inwards. Slow oxidation, or oxygenation (respiration), meantime takes place in the ultimate component cells of the plant, the respired air (containing much carbonic acid) meanwhile diffusing itself and escaping. During daylight the balance of discharge is largely in favour of the oxygen, but at night the balance of discharge is in favour of the carbonic acid.

Thus the plant lives and grows by the exercise of the

Fig. 38.

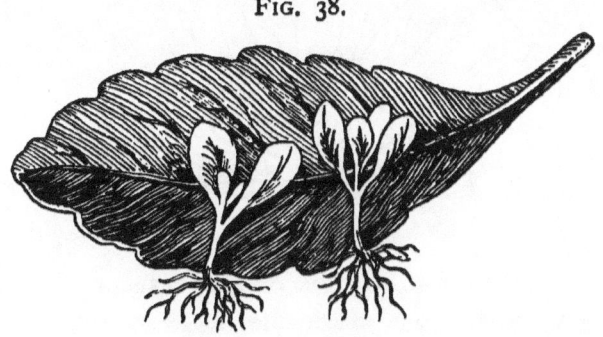

A LEAF OF BRYOPHYLLUM.
Showing two young plants springing from its margin.
(*After Geddes*.)

functions of absorption, circulation, feeding (or alimentation) by its cells, together with respiration and secretion.

Generation, by the aid of the two sexual products (1) the pollen tube, and (2) the embryo cell, provides for the preservation of the race; but this is also abundantly provided for in plants by another process—by a sexless (asexual) process of generation. In many plants aggregations of cells form buds, called *bulbils*, which become detached and then grow into a new individual plant.

Such are the bulbils which form themselves at the roots of the leaves of the tiger-lily. Some ferns give rise to fresh individuals from the surface of their fronds, and a plant named *Bryophyllum calycinum* (Fig. 38) forms buds at the margin of its leaves, from which buds new individuals grow forth. Every one knows how common it is for fresh individual plants, new trees, to grow from "cuttings," and the constancy with which plants will repair injuries and reproduce lost parts after pruning,

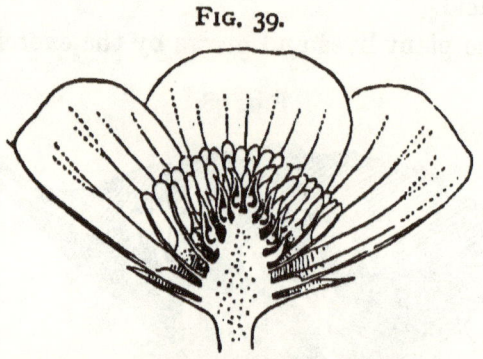

Fig. 39.

VERTICAL SECTION OF A BUTTERCUP.

Showing the floral organs which spring successively from its axis. At the bottom are two of the sepals (of the calyx) cut through. Next come three of the five petals (of the corolla). Then come numerous stamens (each terminating in its anther), which surround six bisected carpels (of the pistil), each showing an ovule contained within it.

is so notorious that the only surprise the reader may probably feel in perusing this statement will be surprise that so well known a fact should be referred to at all.

Every kind of flower and fruit can be understood through the flower of the bean plant, however different from it they may appear to be. Flowers, such as those of the daisy, the dandelion and the thistle, are aggregations of flowers set closely side by side upon a common surface; each separable little group of minute parts in such flowers being a perfect or more or less imperfect flower in itself, on which account such plants are termed *composite*.

In such a flower as the buttercup there is no single body for a pistil, as in the bean, but instead of it there is a number of independent separate parts, each of which is called a *carpel*, and has its stigma on its summit, so that we might say there is a number of pistils, were it not against the custom of botanists so to express the fact. Very often various parts are suppressed, and sometimes a flower may consist merely of a single stamen or a single pistil. Such flowers are, of course, of one sex only, and when a plant has such flowers but bears both kinds—as does the cucumber—it is said to be *monœcious*. In some plants, however, as in the willow, each tree is of one sex, and bears only male or only female flowers. Such a plant is termed *diœcious*.

For all further information about plants the reader is referred to treatises on botany, as our remaining space must be devoted to a brief notice of the structures and functions found amongst animals.

With respect to unicellular animals, little need be said, because their structures and functions so little exceed those of the lowest plants. It must suffice to say that there is a large group called *Rhizopods*, because they can protrude and detract either long or short, thick or filamentary processes of their protoplasmic body-substance. The most beautiful of these are the marine *Radiolaria*, the bodies of which often contain the most wonderfully symmetrical and complex silicious skeletons. There are also the *Foraminifera*, so called because their processes pass out through minute holes in the calcareous shells they form around their most simple bodies. We have already of spoken the *Amœba*,* which protrudes short blunt processes, and so changes its shape in the most protean fashion. All these creatures feed by simply taking in their food at any point of the surface of their bodies and similarly discharging what they cannot digest. There are also the *Flagellata*, which swim about

FIG. 40.

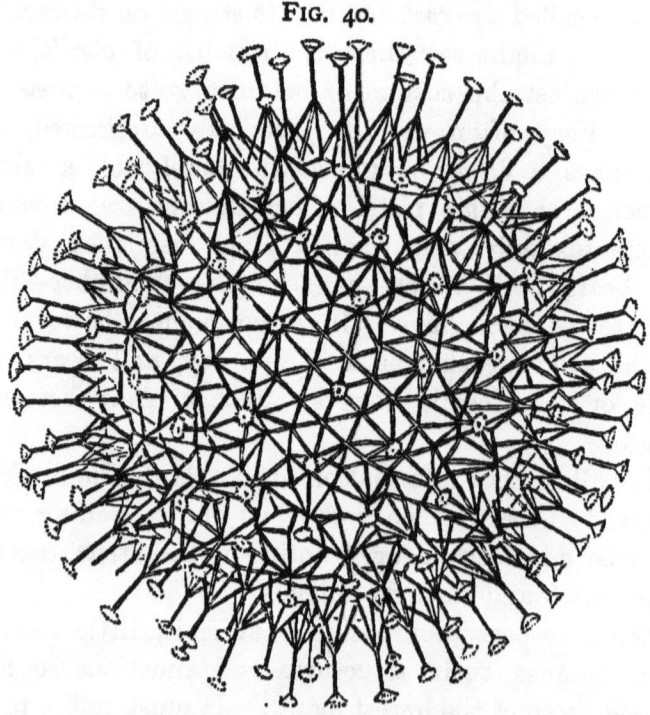

A RADIOLARIAN (*Auloscena mirabilis*) GREATLY MAGNIFIED.

by the aid of one or more vibratile cilia, as does the plant *Protococcus* before described.† Then comes the great group of *Infusoria*, which are slightly more complex in structure and bear bands of vibratile cilia.

Sponges come next, which, very simply organised multicellular organisms, have their cells arranged in

---

* See *ante*, pp. 197 and 198. † See *ante*, p. 200.

FIG. 41.

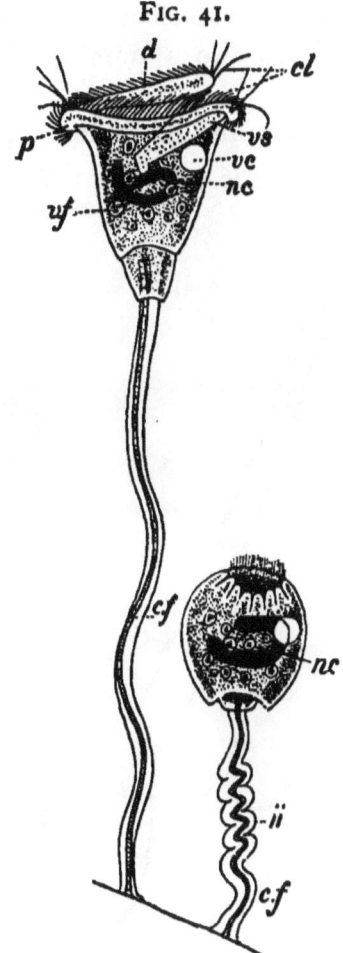

AN INFUSORIAN (*Vorticella*) GREATLY MAGNIFIED.

One individual has its stem contracted, while that of the other is stretched out.

cy Contractile fibre in the stalk not contracted.	*p* One end of the region surrounding the mouth.
cl Cilia.	*rf* Space surrounding a particle of food.
d Upper surface or disc.	
ii Fibre in a contracted state.	*vc* Contractile vesicle or space.
nc Nucleus.	*vs* Vestibule or chamber extending inwards from the mouth.

(*After Howes.*)

two layers, and contain silicious, calcareous, or horny skeletons. They have also a number of inhalent and exhalent apertures.

The zoophytes, or plant-like animals (which include

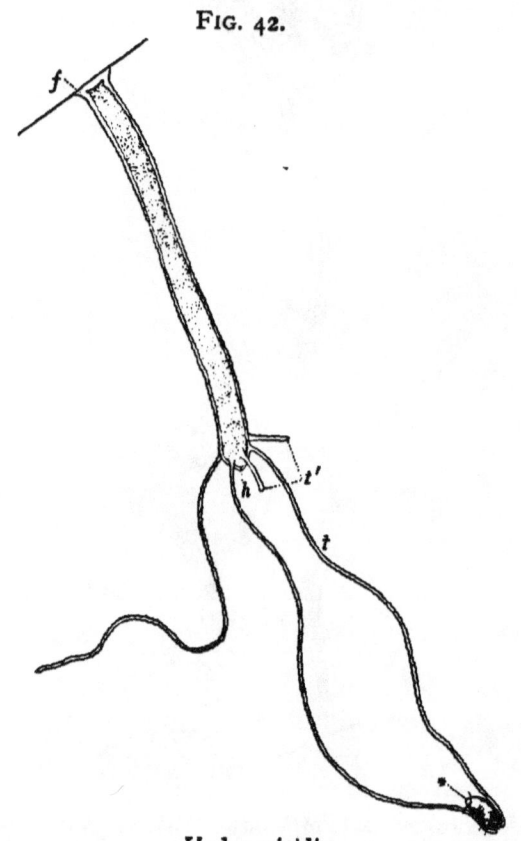

FIG. 42.

Hydra viridis.

  So-called foot.
h Mouth.
t One of the tentacles.

t^1 Tentacles not fully grown.
* Prey seized.

(*After Howes.*)

the coral-forming polyps, which live aggregated together), the jelly fishes, sea anemones, &c., succeed the sponges. The simplest form of the group is the *Hydra*, which consists of a sack with a single aperture surrounded by tentacles, the whole body being made up of two layers of cells, which nevertheless give signs of forming most simple muscle-substance and nerve-substance, tissues to be more distinctly referred to shortly. Another great group consists of the star-fishes, sea-urchins, crinoids, &c., all of which are called *Echinoderms*, and though essentially simple in structure, may consist of a prodigious number of small juxtaposed calcareous parts.

Next may be mentioned the small animals known as wheel-animalcules, and minute creatures which live in compound aggregation, such as the well-known sea-mat *Flustra* (often popularly regarded as a seaweed), which minute creatures are termed *Bryozoa* or *Polyzoa*.

A multitude of creatures are known as *worms*. Such are the many internal parasites, with their allies, and also leeches, earthworms, and aquatic worms.

To these succeed the two great groups of (1) arthropods and (2) molluscs.* Of their structure, however, we will say nothing till we have first briefly described the organisation and functions of one of the highest class of backboned animals—that to which we belong—which class comprises all mammals (or *mammalia*), so called because their females give suck to their young through their mammary glands and breasts.

From this class, which includes all beasts, we will select the cat for consideration, as that will well serve to show how great is the difference between the most complex inorganic body and a highly organised living creature.

The external form of the cat needs no description here. Beneath its skin lies the flesh, and these enclose the bones—skull, backbone, ribs, and limb-bones.

---

* See *post*, pp. 234-238.

Within the trunk is a cavity containing the heart, lungs, kidneys, stomach, intestine, liver, &c. Within the skull and backbone is a mass of white substance—the brain and spinal marrow. Delicate threads of this white substance (nerves), and also tubes of various sizes (vessels), traverse the body in all directions. The various "organs" are, as before said, grouped in "systems,"* and are composed of "tissues." Thus, "fat" is *adipose* tissue, flesh is *muscular* tissue, the outermost layer of the skin is *epithelial* tissue, and its deeper layer is formed of *connective* tissue. Bone is *osseous* tissue, the brain and nerves are *nervous* tissue, gristle is *cartilaginous* tissue, and the blood may be termed *sanguineous* tissue.

The body is reducible to the ultimate chemical elements above enumerated,† but before this extreme reduction to its *ultimate elements* it can be shown to consist of certain complex organic compounds or *proximate elements*, such as *albumen* (the substance of the white of egg) and *gelatine* (the substance of jelly), and others.

The flesh consists of a multitude of delicate threads called "muscular fibres," which are variously aggregated in masses, and so form *muscles*. The heart is a four-chambered muscular organ, the centre of two sets of tubes —*arteries* and *veins*—the extremities of which are connected by minute vessels termed *capillaries*.

A third set of vessels—*lymphatics*—converge from all parts of the body to two large veins. The arteries, veins and heart, are full of blood, and the lymphatics of an almost colourless fluid called *lymph*. These fluids contain a multitude of minute bodies termed "*blood corpuscles*," which are of two kinds, white and red.

---

* See *ante*, p. 187. † See *ante*, p. 192.

The lungs are two very complexly formed air-bags, while a tube—the *windpipe*—descends from the back of the mouth, bifurcates below and then ramifies in each lung, the whole constituting the "respiratory system." The membrane which lines certain parts of this system is coated with cilia *—like those of infusoria, or those by which the protococcus effects its movements.

Each kidney is a rounded mass of minute tubes, which converge to open into a cavity whence a tube descends to the bladder which opens externally by a further canal.

The skeleton is made up of bones and cartilages, and the parts are mostly capable of being moved one upon another by the intervention of the muscles which are attached to them. These movements are facilitated by the shape of the contiguous surfaces of such movable bones, which constitute what are called articulations or "joints."

The essential part of the generative system of the male, consists of very minute tubes which form small bodies analogous to the antherozoids of ferns. They are rounded bodies which move by the aid of a single cilium only, and are called *spermatozoa*.

The essential generative organs of the female do not consist of tubes, but of a peculiar solid substance containing modified cells termed *ova*.

The nervous system consists of an immense multitude of cords, threads, and cells containing a peculiar albuminous fluid. These form the brain, spinal cord, nerves, and small rounded aggregations of nervous substance termed *ganglia*. Nerves proceed from the nervous axis (brain and spinal cord) to all parts of the body, and

---

* See *ante*, p. 200.

certain special nerves pass from the brain to the eye, the ear and nose, and the tongue.

That to the eye expands into a delicate membrane of nervous tissue (of wonderfully complex construction) called the *retina*, upon which an image of external things is projected according to the laws of light,* as in a camera obscura, owing to the various different trans-

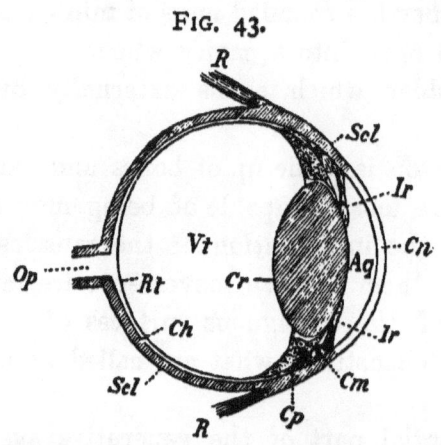

Fig. 43.

VERTICAL SECTION OF A LION'S EYE.

Aq Aqueous humour.
Ch Choroid coat.
Cm Ciliary muscle.
Cn Cornea.
Cp Ciliary processes.
Cr Crystalline lens.

Ir Iris.
Op Optic nerve.
R Muscle of the eyeball.
Rt Retina.
Scl Sclerotic coat.

parent media placed in front of the retina and forming the bulk of the eye-ball.

The true organ of hearing, or internal ear, is a complexly-shaped, membranous bag (containing, and floating in, fluid) on which the auditory nerve ramifies.

The organ of smell consists of minute branches of nerves which proceed from two forward prolongations of

---

* See *ante*, p. 107.

the brain, and ramify in the membrane which lines the back of the nostrils. Similarly the tongue and hinder part of the palate are supplied with tasting or gustatory nerves which also proceed from the brain, though from quite its hinder portion.

The spinal cord gives off nerves on either side symmetrically in pairs, each nerve arising by two roots, one from the front and one from the back of the cord.*

The cat's body possesses what is called *bilateral symmetry*, that is to say there is a close resemblance and correspondence between its right and left sides, though this does not apply to all its internal organs. There is also a *serial* symmetry as, *e.g.*, between the front and the hind leg, between the successive ribs of either side, and between the successive bones which together make up the back-bone.

Such being the structure of the cat, we have next to consider its *physiology*—the functions of its various parts.

We have seen that protoplasm has a power of contraction. This exists greatly intensified in muscular tissue, and all the cat's movements are performed through contractions of its muscles which move the bones to which they are attached, causing them to act like levers of different orders.† Thus, when the cat raises its fore-paw to strike, the paw is the weight, the fulcrum is the lower end of the bone of the upper part of the fore-leg, and the force or motive power is the muscle which is attached to a bone of the lower part of the fore-leg, and which, by its contraction, raises the paw, and so this action is an example of a lever of the third order.

---

* See Fig. 44, p. 226.   † See *ante*, p. 51.

Protoplasmic mobility is seen in the white corpuscles of the blood, which can change their shape as does an *amœba*, which indeed they closely resemble.

It is also shown by the action of the cilia of the respiratory system, which move harmoniously like the stalks of a field of corn under a strong wind. The result of this is that they propel forwards any small body which may be upon them, and thus the breathing organs become liberated from various matters which are so borne upwards towards the mouth.

The great process of nourishing the body is effected through the mechanical division of food by the cat's teeth and its solution by the juices of the spittle-glands, the stomach, and the alimentary canal, into which canal the secretions of the liver and pancreas are poured. These juices so act on the food as to change many of its component parts from an insoluble into an easily soluble state — namely* from "colloids" into "crystalloids," but, as before said, the final process consists of what is called *assimilation*, or the transformation of matter external to the most intimate substance of the cat's body, into that very substance—the change of the food it eats into the cat itself.

But besides nutriment, the animal requires that its body should be kept at a certain temperature, and this is effected by that continuous process of slow combustion †—oxygenation—before mentioned.

But nutrition could not be effected were not fresh nutritive material conveyed all over the body to replace wear and tear; and it is so conveyed by the circulating system, the blood exuding from the finest capillary vessels, to reach the ultimate cells and structures of the body.

---

* See *ante*, p. 145.   † See *ante*, p. 194.

But the blood after being thus impoverished, itself requires replenishment, and this it obtains from the nutritive material gathered from the alimentary canal by the lymphatics, and subsequently poured into the blood.

Of all the functions of the body, that of *respiration* is the most conspicuously necessary for the maintenance of life, since the separate life of the cat begins with an act of inspiration, while it is with an act of expiration that it ceases. A short interruption of the process necessarily results in death. In breathing, the air is taken down into the lungs, and is thence again expelled much poorer in oxygen but containing a much increased quantity of carbonic acid, because the blood which comes to the lungs from all parts of the cat's frame liberates into them the excess of carbonic acid it has obtained from the ultimate substance of the body and, in exchange, takes from the air in the lungs, the oxygen requisite to supply that ultimate substance with the oxygen it needs.

Thus, in such an animal as the cat, there is both an internal and an external process of respiration. The former consists of the gaseous exchange which takes place in the ultimate particles of the body. The latter is the gaseous exchange which takes place in the lungs themselves.

Closely connected with respiration and nutrition is the process named *secretion*, which is the special function of such organs as are called *glands*. Two most important organs of the cat's body are the *kidneys*, which secrete and remove from the blood certain effete and deleterious nitrogenous substances. The salivary glands, liver and pancreas, pour their respective secretions (spittle, bile, and pancreatic juice) into the alimentary canal, the walls of which are replete with small glands, as are those of the stomach, which latter secrete *gastric juice*.

The generative function is a special modification and form of "growth," while growth is a sort of self-generation. This is specially perceptible when any part which has been destroyed is reformed, as when a broken bone is repaired. Then the two broken edges become softened, and a substance is secreted between them which is at first jelly-like, then gristle-like, and at last bony. In generation, impregnation is effected by the junction of the spermatozoon with the ovum, which is a process essentially similar to that of the union of antherozoid and pollen tube, with the female cells of the fern and of the bean plant. Immediately after impregnation very curious changes ensue, which cannot here be described, the reader being referred for their explanation to treatises on embryology.* Here it must suffice to say that the first germ of the future animal appears in the shape of a minute rounded mass of protoplasm, which divides and divides itself again and again till three layers of cells are formed, whence, by degrees, all the varied tissues and all the complex parts which constitute the kitten are gradually but rapidly built up. The building up of the kitten, as it exists at birth, goes on nevertheless in a roundabout fashion, various structures being for a time formed which resemble conditions that are permanent in lower animals, but which subsequently disappear or become much modified.

The functions of the cat's nervous system merit some special consideration, since without its aid none of its other bodily activities could be carried on. Its functions also present the most extreme contrast yet met with to

---

* And to my work on the cat, p. 317.

each and all of the powers possessed by inorganic bodies. The processes of growth and generation are different indeed from the activities of non-living bodies, but far more divergent is the power of feeling, of perceiving by the senses surrounding objects, and of regulating bodily actions in conformity thereto. *Sensitivity* is the special attribute of nervous tissue, and each creature's nervous system is thus an organ of intervention between it and the world around it.

Sensation can be absolutely known only to the being that experiences it; nevertheless it would be in the highest degree absurd not to be certain that a cat feels, sees, hears, smells, and tastes, and that it possesses feelings of pleasure and pain, with propensions, desires, and emotions of affection, fear, animosity, &c. But different parts of the nervous system have different functions, as is the case with different parts of one portion of it. Thus part of the spinal cord transmits an influence upwards to the brain, resulting in sensation, while another part transmits an influence downwards from the brain, resulting in movement. The ascending sensitive influence passes to the cord through the posterior roots of the spinal nerves into which enter the nervous fibres terminating in the skin, and which are affected when the skin is touched. The descending motive influence passes from the cord through the anterior roots of the spinal nerves, which ramify and terminate in the muscles where they produce motion.

It is not, however, necessary that such influence should actually ascend to, or descend from, the brain in order that responsive motions should take place, although it is certain (from observations made on accidentally injured men and purposely mutilated animals) that the brain must act in order to give rise to sensation.

When the influence originated by touching, pricking, or burning some part—*e.g.*, some part of the leg—is prevented by any cause from reaching the brain, while none the less some appropriate action follows—*e.g.*, the withdrawal of a foot from a hot iron, without the occurrence of any sensation—such response is called *reflex action*. The unfelt influence travelling upwards and inwards is supposed, on reaching the spinal cord through the posterior roots of its nerves, to be there automatically reflected outwards, through their anterior roots,

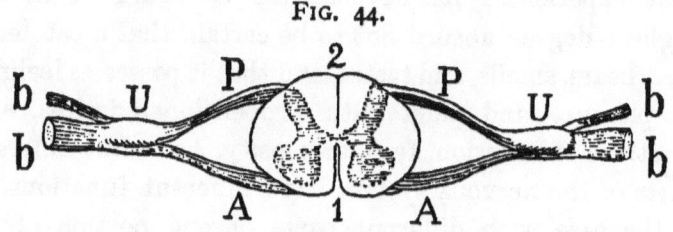

FIG. 44.

TRANSVERSE SECTION OF THE SPINAL CORD AND ROOTS OF THE SPINAL NERVES OF A CAT.

A Anterior roots.
P Posterior roots.
U Compound nerve formed by the junction of these roots.
b Branches given forth from the united nerve.

1 Anterior (or ventral) median fissure of spinal cord.
2 Posterior (or dorsal) median fissure.

to the nervous fibres which pass to the muscles and excite motion in them.

But a response quite independent of volition may also take place when feeling is in no way impaired. Thus, if a small object be placed sufficiently far back in the mouth, the muscular act of swallowing will be performed automatically.

It is evident, as before said, that the cat can feel pleasure and pain, and can experience a variety of definite sensations (of sight, sound, odour, &c.), which the creature can so employ as to have a practical sense-perception (through its different and combined senses) of different objects which it also practically distinguishes from itself.

Feelings experienced, successively, tend to become associated together. Thus it is when some definite past feeling is revived, an imagination of feelings, and groups of feelings, previously associated with that past definite feeling, will again present themselves to the cat's imagination, and create expectant feelings as well as render sense-perceptions more distinct and significant.

It is impossible to doubt that a cat can at the same time see, feel, smell, and taste a mouse it has caught, as also that it can hear its cry while seeing it and clawing it, before killing it. There must, therefore, be some common centre where these influences are simultaneously received. There is no reason to suppose the cat can know that it exists and that it is not the mouse, but it evidently possesses a feeling, however vague, of this distinction. This feeling of self-identity and distinctness from other things, is to be distinguished as a feeling of *consentience*. That the cat possesses a power of retaining and reproducing groups of feelings which have before been excited in it by external objects, can hardly be doubted, since dogs give sometimes such plain signs that they dream, and dreams are the reproduction, by imagination, of sense-perceptions previously experienced. The cat has also the power of associating effects of past sensations with present ones in such a way that the occurrence of one will excite the other. We cannot doubt that the sound of a gnawing mouse, or a perception of its odour, will excite in the cat's imagination an image of a mouse, and this may lead it to intensify the exercise of its senses and so simulate what we know as "attention." This implies that the animal possesses a certain power of *memory*, though no one supposes that the cat notices its own recollections, as such, or ever sets itself to try and recall something temporarily forgotten. Similarly, though the animal does not note that objects are of certain shapes and sizes, that they are few or many, that they are in a particular place or that they move and so change their relative positions, nevertheless, its faculties enable it to practically respond to the different feelings induced by all such external relations which exist between it and surrounding things. The cat acts differently according as only one mouse or two mice are present, and according as a mouse is still or is running away, and it regulates its movements of pursuit according to the changing relations which the mouse's flight gives rise to, between the mouse itself and the pieces of furniture in a room through which it tries to escape. The cat may also experience surprise, feel puzzled and have its attention strongly excited, without being aware of those experiences as such. Similarly it will with amazing rapidity take means to effect a desired end,* sometimes by jumping to undo the latch of a door; but it does this without recognising that it is, in fact, taking means to effect an end. By such movements it really acts as a cause producing an effect, but it does not regard itself as a cause or recognise effects produced by it as being what they are.

Again the experience of some slight sensations which have often before occurred as preliminaries to other vivid ones, and have so become associated therewith—as a jingle of cups preliminary to the experience of a saucer

---

* See "The Cat," pp. 365-371.

full of milk—will give rise to an expectant feeling* which is often subsequently gratified; but this does not imply that the cat mutely says to itself, " Sounds of cup jingling are probable preliminaries to milk tasting. The sounds I hear are cup jinglings, therefore the sounds I hear are probable preliminaries to milk tasting."

Finally, the cat has a certain vocal language and a language of gesture. That is to say, it emits different sounds according to the feelings it experiences, mewing, purring and spitting as the case may be. The gesture language of a cat to a beloved mistress is often very expressive of attachment, while the gestures it will exhibit to a threatening dog are unmistakable indeed. The cat by the sounds it emits, or the gestures it makes, has the power of so giving rise to corresponding feelings in other creatures that the latter can often practically understand what it means, as the cat can understand, to the same extent, the meaning of the sounds and gestures of a threatening dog. Such language may therefore be called a *language of feeling* or " emotion."

The cat has also the power of acquiring certain *habits*. Now a habit is a curious and interesting thing. A "habit" is not formed by repeating actions, though it may be strengthened by them. If an act performed only once had not in it some power of generating a "habit," a thousand repetitions would not generate it. Most animals (certainly such an animal as the cat) have a natural tendency to activity—a positive want of it. An animal's powers also tend to increase with activity (within limits) and diminish with too prolonged repose. Habit is then the determination in one direction of a previously vague tendency to activity.

---

* See *ante*, p. 227.

But the cat also possesses *instincts*. The action by which the kitten first sucks the nipple and swallows the thence extracted nutriment are *instinctive* actions. They are necessary, definite actions which have never been learned but are performed prior to all experience.

Every one knows that kittens more or less resemble one or other of their parents. This transmission of parental characteristics is called *heredity*, and is evidently a property of the parents which transmit their likeness. Unusual peculiarities, such as additional toes

FIG. 45.

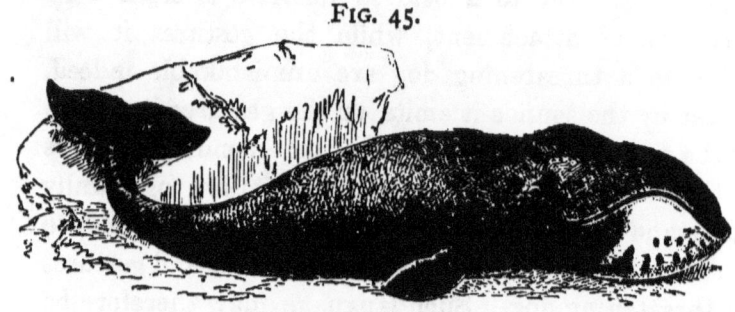

THE RIGHT WHALE.

and claws, are characters which also tend to be inherited.

Such is the cat, and a variety of beasts—lions, tigers, leopards, lynxes, &c.—are formed almost entirely in the same manner, save as regards size. Other beasts differ from it in greater or lesser degrees, till we come to such forms as bats and monkeys, whales and porpoises. A monkey or ape is formed much like a cat, but the proportions and shape of the paws and toes are very different, and the same is the case, though to a less degree, with the limbs. The teeth, also, are very different. They are suitable for eating fruits, not for cutting and dividing flesh. There are various other orders of beasts, all of which resemble the cat more closely than they do any bird, reptile, or fish. In birds, we meet with very beautiful structures—feathers—which are found in no animal which is not a bird; and large feathers are almost always present in the wings and tail. There are nearly 11,000 different kinds of birds, but they are all formed much on the same model,

FIG. 46.

THE RHINOCEROS VIPER.

the difference, *e.g.*, between a nightingale and a goose, being slight indeed, compared with those which exist between a horse and a squirrel amongst beasts, or between a tortoise and a snake in the great group of reptiles.

As we all know, parrots and some other birds will learn to articulate words and sentences with great distinctness, without, of course, understanding the signification of the words to which they give utterance, and which they evidently take pleasure in repeating.

All birds have two pairs of limbs, but many reptiles have none, and serpents creep over the ground by the aid of their numerous and very movable ribs, while their

FIG. 47.

THE DEVELOPMENT OF THE FROG FROM THE TADPOLE.

1 to 7, successive stages. 2, with external gills, which have disappeared in 3.

excessively distensible jaws enable them to swallow creatures of very large size, compared with that of their own head and neck.

Reptiles are not normally able to raise the temperature of their bodies above that of the atmosphere they live in, but birds are, as before said, very hot-blooded, their warmth being promoted by the fact that air passes from their lungs into different parts of their bodies, sometimes even into the substance of almost all their bones.

Frogs and toads are animals which, in their young

FIG. 47.

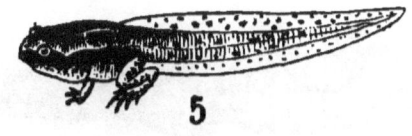

THE DEVELOPMENT OF THE FROG FROM THE TADPOLE.

4 and 5 show the limbs growing out; 6 and 7 the tail very nearly and quite absorbed.

condition, possess a power which has not yet been noticed. Their young, called tadpoles, are aquatic and do not at first breathe at all by lungs, but by the intervention of delicate processes of skin which are placed on either side of the neck and which float in the water,

Such structures are termed *gills*, and in them the blood exchanges its carbonic acid for oxygen, as it does in the lungs of the cat and other beasts. It is not that the water is dissolved into oxygen and hydrogen. The oxygen absorbed is gained from air which is mixed up [*] in the water. When this air has been expelled by boiling the water, no animal with gills can any longer live in it. Adult frogs and toads have no gills but lungs, by which they breathe, as is the case with almost all their tailed relatives, the efts, and as all the higher animals do.

The immense group of fishes all breathe by gills,

Fig. 48.

THE EFT AMBLYSTOMA.

though a few kinds also possess an apparatus for breathing air which is more or less comparable with a lung.

Beasts, birds, reptiles, frogs, and their allies, with fishes, make up the great group (class) of vertebrate, or back-boned, animals.

We can here only refer to two other great sub-kingdoms, referring the reader for all else to works devoted to zoology.

The first of these is far richer in the number of kinds it contains than are all the other classes of animals taken together. It is the sub-kingdom *Arthropoda*,

---

[*] See *ante*, p. 153.

which embraces all insects, hundred and thousand legs, scorpions, spiders, tics, and mites, all lobsters, crayfish, crab and shrimp-like creatures. As an example, we may select the crayfish, the body of which is evidently in part composed of a longitudinal series of similar segments, while numerous pairs of lateral appendages successively appear (from before backwards) as feelers, jaws, claws, feet, and swimming paddles.

There are two conspicuous eyes, each borne on a stalk. A nervous system, composed of longitudinal bands and

FIG. 49.

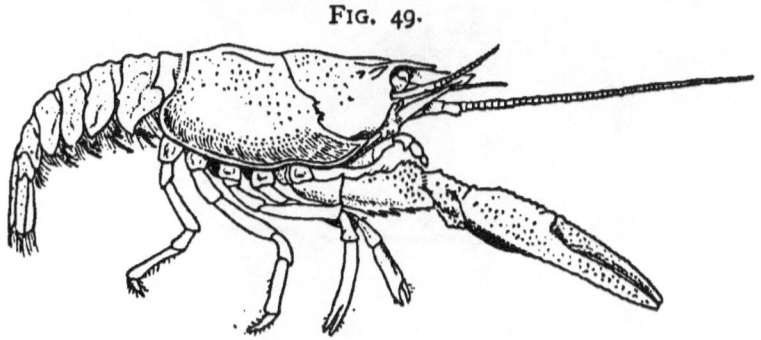

CRAYFISH (*Actacus fluviatilis*).

ganglia, runs along the body inside its lower or ventral surface, whereas in vertebrates it is situated in the dorsal region or back. It is there, in the crayfish, that the heart, a single-chambered organ, is placed, whence blood-vessels proceed, the blood being purified by the aid of gills which project upwards from the bases of the legs. In this form of body, serial symmetry is carried to a far greater extent than in vertebrate animals, while bilateral symmetry is no less obvious.

The class of insects are remarkable for their power of flight; but their wings have no resemblance, save as regards the function they perform, to the wings of vertebrate animals, whether bats or birds. Insects breathe air, but not by lungs. Instead of one windpipe, they have many, which, opening on different parts of the external surface of the body, ramify inwards, and carry air (for respiration) to all parts of the frame. Scorpions, however, do not thus breathe, but by means of a series of small sacs, which open in the under surface of the body and admit air within them. The jaws of arthropods are quite different from those of vertebrates, and

FIG. 50.

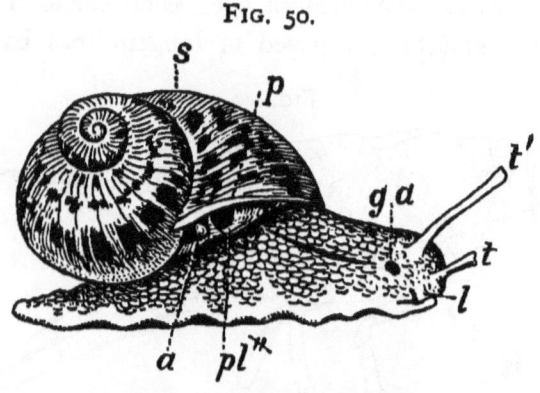

THE SNAIL (*Helix pomatia*).

a	The hinder termination of the intestine.	*p*	Shell margin.
		pl''	Respiratory aperture.
ga	Generative aperture.	*t'*	So-called olfactory tentacle.
l	Lip.	*t*	Eye tentacle.
s	Shell.		

(*After Howes.*)

bite laterally, and neither from above, downwards, nor from before backwards.

The other sub-kingdom to be noticed here, is that of the molluscs (*Mollusca*), and embraces all snails, whelks, cuttle-fishes, oysters, mussels, &c. Almost all of them breathe in water by means of gills, but a few, as the snails, perform aërial respiration by the aid of a small sac which admits air within it.

The snail is an organism which contrasts greatly with

an arthropod. Instead of a succession of jointed limbs, the animal moves upon a single, elongated, muscular expansion, called the *foot*. Indeed, serial symmetry is almost abolished, the viscera being arranged in a spirally coiled projection upwards from the foot, and this projection is protected by a similarly coiled calcareous shell. There is a well-developed head, with two pairs of retractile processes, the larger of which bear eyes. On the right side of the animal, close to the anterior edge of the shell, is a large aperture which is that of the breathing sac. There are no jaws, like those either of a vertebrate or an arthropod, but within the mouth there is a crescent-shaped plate above, and a cartilaginous cushion or pad, bearing teeth, below.

The nervous system is in the form of a ring of nervous tissue round the gullet with ganglia, whence nerves proceed to a third pair of ganglia.

The cuttlefish is like a snail devoid of an external shell and with the body not spirally coiled, while the margins of the foot are drawn out into long sucker-bearing arms, the eyes being sessile. It is furnished with a pair of gills formed somewhat like those of the lobster, and the main ganglia of its nervous system are protected by a sort of cartilaginous skull.

The oyster, mussel, and their allies mainly resemble

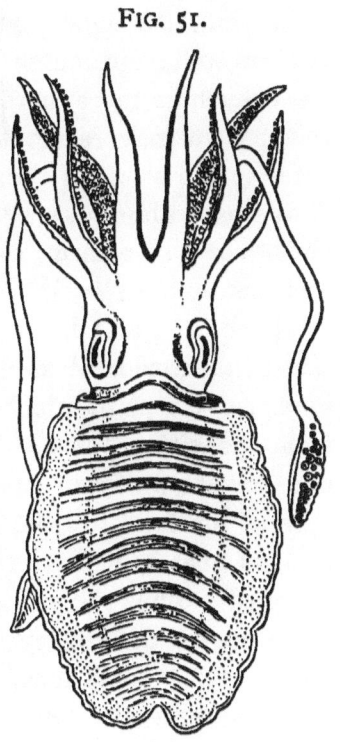

FIG. 51.

CUTTLEFISH (*Sepia*).

the snail in structure, but with the following exceptions: there is no head and no air sac, but there are plate-like gills, and the shell, instead of being a single, spirally coiled cone, consists of two lateral halves, united dorsally by a hinge, and bearing to the creature between them, somewhat the relation of the two sides of a frock coat to the man who wears it.

Having, it is hoped, now said enough to stimulate our readers to have recourse to works on zoology in order to pass beyond a mere introduction to the elements of that science, we may revert to general considerations which refer to the whole of the organic or living world.

We have already seen* what are the properties of that substance, protoplasm, which is common to all animals and plants, but the simplest facts of physiology suffice to establish the great distinction between the living and the non-living world. A seed, under suitable conditions, will give rise to a plant which will again produce a seed, and from the kitten there will similarly be produced a cat, and thence a kitten once more. So the changes of organic life tend to recur in *cycles*—the necessary conditions of heat, moisture, and gaseous material, &c., being supplied. Thus the existence of an innate tendency to go through a definite cycle of changes when exposed to certain fixed conditions, forms a distinction, not only between mineral substances and living organic bodies, but also between living organisms and those which have died. The latter will go through changes indeed, but not a *cycle* of changes—they never return to the point whence they set out.

Inorganic substances tend simply to persist as they are, and have no definite relations either to the past or

---

* See *ante*, pp. 192-197.

to the future. But every living creature, at every stage of its existence, regards both the past and the future, and thus lives continually in a definite relation to both of these, as well as to the present. It has, therefore, under the conditions necessary for life, a definite spontaneous activity of its own. An inorganic body may be one kind of substance, but it is only a living organism which can be called an *individual*. As yet science has not afforded us any means whereby we may give origin to life— whereby we may change any non-living substance into a being instinct with life.

Plants possess the power of performing all the functions essential to continued life—namely, alimentation, circulation, respiration, secretion and generation. They give no evidence, however, of possessing any true power of sensation. The movements of the sensitive plant, the sun-dew, Venus's fly-trap, &c., are very wonderful, but the most careful examination of those plants has not discovered any nervous tissue in them. As this is the essential and only organ of feeling which organisms are known to possess, its absence in plants justifies our denying them the power of sensation. To say that our ordinary domestic plants—our potatoes, and our cabbages —really feel, would be absurd in the eyes of men of common sense.

The movements of plants are also effected in a different manner from those of animals. In the latter it takes place by the aid of muscular tissue, a substance which has not yet been found in any plant, and it is a rule, gathered from long experience, that "structure" and "function," in organisms, vary together.

As already pointed out, no animal can live without feeding, directly or indirectly, upon plants, since they alone possess the power of building up organic matter directly from the inorganic world. With the exception of fungi and some parasitic plants (such as the dodder), the whole vegetable world is continually engaged, during the hours of daylight, in tearing from the atmosphere its carbon, and in absorbing moisture in order to build up substances capable of life.

The repair of injuries and reproduction of lost parts take place in lower animals to a much greater extent than it does in the one selected as our type—the cat. The tails of lizards, the limbs of efts, and the legs and claws of lobsters, if broken off, will be reproduced, and some aquatic worms have been cut into as many as twenty-five parts, with the result that each separated part has grown into a whole. Some polyp-animals form buds (like those before spoken of as being produced by tiger lilies), which will often become detached, and then grow up into new individuals, like those plants which will give forth "suckers" and then separate. The common bramble will attach itself to the ground by the end of a "shoot," rootlets coming to take the place of the incipient leaves of its terminal bud, and so a new stem is formed.

Thus "growth" is "continuous reproduction," and "reproduction" is a form of growth which may be "continuous" or "discontinuous." Continuous reproduction occurs in animals as well as in plants, and thus it is that many coral animals grow up as arborescent structures, or into large masses leading to the formation of reefs and islands. Discontinuous growth may occur in certain worms, which habitually divide themselves and so multiply, and multitudes of Protozoa, as before said, multiply by spontaneous fission.

The circuitous course of development of the embryo, before mentioned as taking place in the kitten, is pursued to a greater or less extent in the development of almost all animals.

The embryos of higher animals for the most part transitorily resemble, in their general features, the structure of other animals lower in the scale. The series of forms, also, through which the embryo of a higher animal passes in its development (or *ontogeny*), successively resembles, in a general way, a series of adult forms of animals lower in the scale of life.

Thus the heart of a cat is at first but a single tube, as it permanently remains in the ascidians or sea squirts. The cat's brain consists in its earliest stages of a series of simple vesicles, roughly like the brain of that lowly fish, the lamprey. In a more advanced stage, the embryo of the cat is plainly the embryo of a beast—not of a fish— and later on it is plainly that of some beast of prey.

Most remarkably obvious are those changes which take place when the embryo is a free active creature during its development.

Thus the young of the frog is, as before said, a tadpole, and so the frog in its development is said to undergo a *metamorphosis*. No less marked is that change which most insects undergo, and which is so well seen in butterflies and moths, which are first actively feeding grubs, then quiescent crysalides, and finally reproductive, usually winged, adults. These stages are known in zoology as (1) the *larva*; (2) the *pupa*; and (3) the *imago* or the perfect and mature insect.

Thus the whole of the living organic world begins, as it were, from a common unicellular starting-point, whence the two kingdoms of organic life may be said to diverge. The animal kingdom advances in complexity from a structure resembling a double-walled sack, with a permanent digestive cavity, and possessing nervous and muscular tissue—such a condition we find in the *hydra*. The vegetable kingdom advances in complexity in a quite diverse mode, building up a variously branching axis with foliar organs (modified leaves), but always devoid of any alimentary cavity and any form of muscular or nervous tissue.

The functions which are peculiar to the higher organisms, and are exhibited by all living creatures which possess nervous and muscular tissue, are (as has been before said) those of movement and feeling. These two functions are distinguished as *the functions of animal life*, in contradistinction to the functions of nutrition and reproduction, which, being possessed by all plants, as well as animals, are termed the *vegetative functions*. That the animals with which we are most familiar have feelings and emotions, and that we can, to a considerable extent, tell what these are, hardly any one will be disposed to deny. As to lower animals, the complex social economy of bees is a matter of common knowledge. Ants display a complete and yet more complex political organisation. Some have soldiers which capture slaves, while other kinds will retain other insects (*Aphides* *) captive to serve a purpose analogous to that of our milk-giving cattle.

We have already spoken of the vocal and gesture language of the cat. Pointers and setters will make certain facts known by their gestures; the songs of birds have meanings practically understood by their fellows, while parrots and jackdaws can learn to articulate whole sentences.

As to the mental faculties of the higher animals, they

---

* The small slow-moving green flies so common on rose trees and pelargoniums.

may be astonished, but they have no recollection of *being* astonished. They can distinguish an artificial object from the natural object which it imitates, but they do not understand the artificial character, as such, of the former. A dog may fear another dog which is stronger and fiercer, but it will have no idea of "courage" and "fierceness." Many animals, even insects, will distinguish clearly between differently coloured objects—the white from the blue, the red from the yellow—but no animal gives us evidence that it knows "whiteness" or "blueness," and still less that it knows what "colour" is. Some animals also have feelings of sympathy, companionship, regretful feelings, feelings of shame, &c., but we have no ground for supposing they understand the conceptions "ought" and "duty." Animals generally possess the faculty of forming "habits," but the instinctive powers of many of them are much greater than those of the cat. Chickens, two minutes after they leave the egg, will follow with their eyes the movements of crawling insects, and peck at them, judging distance and direction with almost infallible accuracy. They will also instinctively appreciate sounds, readily running towards a hen hidden in a box when they hear her "call." Some birds will feign lameness in order to draw off attention from their eggs and young, and birds of the first year will readily migrate to avoid a cold of which they can have no knowledge. But it is insects which possess the most remarkable instincts, such as those of the carpenter bee, the wasp sphex, and the Emperor moth, and many others, for a description of which the reader is referred to works on zoology, and especially to those on *entomology*—the natural history of insects. Such phenomena make it clear that insects will make elaborate arrangements for a progeny they can never see, and the habits and food of which differ widely from those of the parent since its larval condition. We cannot think, however, that the insect possesses any recollection of that condition so that its parental actions are guided thereby.

As we said in the beginning of the chapter, organisms have definite relations (1) to time; (2) to space; and (3) to one another.

As to *time*, it is widely known that some animals have become extinct. Thus the wolf has disappeared from England since the days of Henry VIII., while the bustard has ceased to exist, although eighty years ago it wandered over the South Downs and Salisbury Plain. Similarly, plants once common in certain places have since vanished, as the many peculiar plants of St. Helena have been almost entirely destroyed by the rabbits and goats introduced into that island.

The evidences we possess of past organic life is afforded us by the five kinds of fossils before described.* This record is an exceedingly imperfect one, remains of animals and plants having been only here and there exceptionally preserved by some favouring accidents, and often in a very fragmentary manner. The study of these organic remains constitutes the science of *Palæontology*, and we must refer our readers to treatises on that science for further information. Here we will only say that in the primary rocks have been found many remains of echinoderms, molluscs, arthropods and fishes.

In the secondary strata we find evidences that great numbers of huge reptiles existed, some grazing or feeding on trees, and others of most predacious habits.

---

* See *ante*, p. 170.

Then our present whales and porpoises were anticipated by gigantic marine reptiles, *ichthyosauri* and *plesiosauri*, while numerous flying reptilian forms, small and large, flitted through the air as bats do now. That time may well be called the age of reptiles. Nevertheless, small beasts had then already begun to make their appearance.

In early tertiary times creatures existed which more closely resemble forms now living, while in the later tertiaries are entombed the remains of organisms more

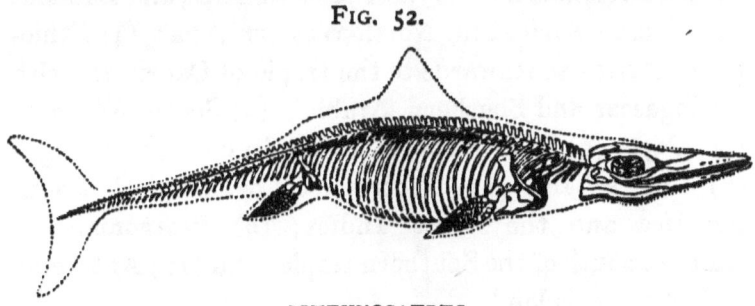

FIG. 52.

ICHTHYOSAURUS.

Showing the outline of the dorsal and tail fins, the existence of which has been recently discovered.

and more like those now living as we examine strata later in date.

That living organisms have definite relations to *space* has been already noted and some examples given.

Each large portion of the earth's surface has, in fact, its special plant population or *flora*, as it has its special animal population or *fauna*. For details the student is referred to works on organic geography. Nevertheless it may here be noted that (1) South America, (2) Africa south of the Sahara, (3) Australia and (4) India with its Archipelago, have each an interesting and peculiar fauna; as also that (5) North

America on the one hand, and (6) Europe, with North Africa and Asia, on the other, are similarly, though less strikingly, characterised. Various species which now inhabit one or other of those areas closely resemble certain tertiary fossils also found therein.

The botanical regions into which the world is divisible are ten in number. (1) Arctic; (2) Boreal (or Europe, Asia and America, from the Arctic circles to the Pyrenees, the Alps, the Balkans, the Himalayas and North America to the tropic of Cancer); (3) Caucasia, or the shores of the Mediterranean up to the Pyrenees, Alps, and Balkans, with North Africa and North-Western Asia; (4) Ethiopia or Africa southwards to the tropic of Capricorn, with Madagascar and Southern Arabia; (5) South Africa to the Cape; (6) Indo-Malayan, or the Indian Archipelago; (7) Australian; (8) Neotropical, or tropical South America and the West Indies; (9) Patagonian, or America south of the Southern tropic; and (10) Antarctic, or Kerguelen's land.

The geographical range of living creatures, even of the same class, is often most unequal. Thus the flame-bearing humming-bird is confined to the crater of the extinct volcano Chiriqui, in America, while the crow ranges over almost the whole world except South America. Different animals and plants are obviously influenced as to their extent of range by their different requirements as to heat, light, moisture, &c.

Beside geographical distribution, living organisms have a vertical or *Bathymetrical* distribution. Thus in the tropics, palm, bananas, &c., grow luxuriantly in the lowlands. At a moderate elevation they give place to evergreens, then these to a belt of deciduous trees, then we find only shrubs, grasses, Alpine plants, and mosses. The camel is an animal of the plains, but the llama will ascend to 18,000 feet in the Andes. The viper is found in the Alps at 5000 feet above the sea, and in South America the condor will soar to more than 22,000 feet above the sea-level.

The ocean has different inhabitants at different depths, and there seems to be no depth-limit to life, especially to animal life. At a depth of 2000 fathoms the ocean fauna present much richness and variety.

As to the inter-relations of living organisms, one great organic inter-relation, already noticed, underlies all others—namely that by which oxygen is set free by plants to be made use of in respiration and so recombined with carbon—a process which has been called the "circulation of the elements."

The phenomena of parasitism constitute a very common relation between organisms, parasites being generally more or less inimical to their hosts.

Besides the evident relation of "enemies" and "rivals," organisms may also directly or indirectly benefit other creatures.

Thus ants and the bull's-horn acacia benefit each other directly. The plant furnishes food and lodging (in special cavities) to the insects, which in turn are ready to rush out and bite furiously any animals which attempt to injure the tree. Organisms may be benefited indirectly by others which destroy the enemies or rivals of the former. Again, services may be rendered in very curious ways. Thus barnacles, which fix themselves on whales, are provided with an extra amount of food by being so carried about. A small fish has also been found to live within the interior of a sea-anemone, feeding on portions of the latter's food. Birds disseminate seeds which they have swallowed, and sometimes do so by carrying them—or the eggs of small animals—

in the mud which may adhere to their feet. Many curious arrangements exist by which the access of insects to flowers—which, as before said, they accidentally fertilise by dusting the stigma with pollen—is facilitated, and others whereby the approach of noxious creatures is prevented.

Plants, the pollen of which is only carried accidentally by the wind, may have each of their pollen-grains furnished with a membranous expansion which facilitates its carriage.

There is sometimes a curious resemblance between different creatures, which goes by the name of *mimicry*. Familiar examples of mimicry are clear-winged moths, which may be readily mistaken for bees. One of the most perfect examples of mimicry is displayed by the insect called the "walking leaf," which in form and colour so closely resembles a leaf that it is difficult to find it when amongst real leaves, and thus it escapes its enemies.

Other creatures, called "bamboo-insects," resemble a stick of bamboo, and this the more because they have the habit of hanging with their long legs stretched out unsymmetrically.

Certain African species of those plants which are called euphorbias so greatly resemble some American cacti, that it is difficult to believe, when out of flower, that they are not close allies, instead of being plants which belong to widely different groups.

Very many animals partake of the colour of their usual surroundings. Thus, desert snakes and lizards are generally sand-coloured, while those which inhabit trees are green.

Actual changes of colour sometimes preserve this harmony. Thus the ptarmigan, the variable hare, the ermine and the Arctic fox become white in winter and so match with the snow.

Sometimes creatures will change colour readily, as is the case with the chameleon and some insects; modifications of form and colour often attend the advent of the

FIG. 53.

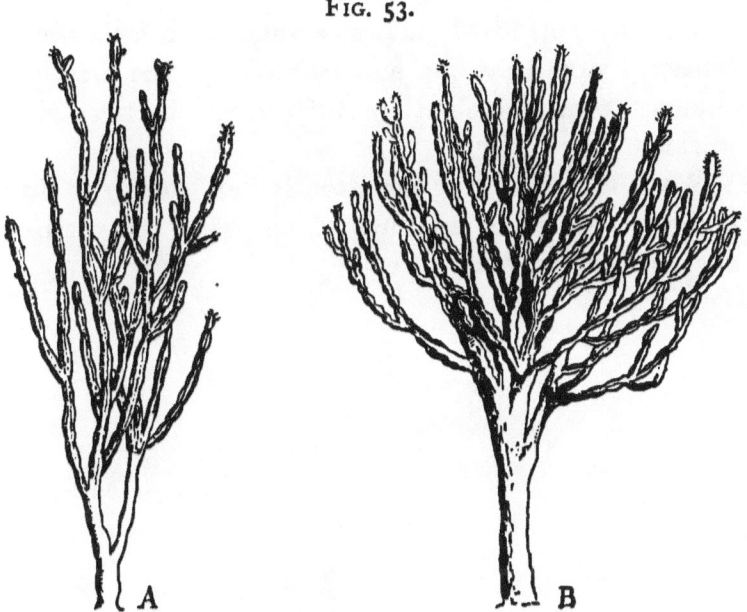

TWO PLANTS, VERY DIFFERENT IN NATURE, WHICH GREATLY RESEMBLE EACH OTHER EXTERNALLY.

A A cactus (*Cereus*) from Brazil.
B A spurge (*Euphorbia*) from Africa.

breeding season, with vocal manifestations, such as of singing birds, croaking frogs, and rutting stags.

The environment of organisms will also affect them in many other ways, besides their colours. Thus the use of soft food seems to have diminished the jaws of both civilised men and their pet dogs. English oysters transported to the Mediterranean are said to alter their mode of growth rapidly, so as to resemble the kind natural to that sea. Setters bred at Delhi and cats at Mombas in Africa have been known to undergo quickly, or in a generation, considerable modifications, while twenty kinds of American trees all differ from their nearest European allies in a similar manner—less toothed leaves, fewer branchlets, &c.

We have now briefly indicated some main facts which concern living creatures, and pointed out the principal circumstances whereby they differ from the non-living world.

We must next direct our attention exclusively to that highest of all living creatures which people this earth—our own species, Man.

CHAPTER VII

MAN

THE human body is formed on the same fundamental type of structure as is the body of every beast, and therefore that of the cat hereinbefore briefly described;* but he resembles above all the ape in structure, so that, thus considered, apes and men may be said to stand together on a sort of zoological island, widely separated from all other animals. The difference in structure between an ape and every other animal is very much greater than that which exists between man's body and that of any ape.

All races of men are very similarly formed. Differences exist, as everybody knows, with respect to the colour of the skin, the form and abundance of the hair, the prominence of the jaws, and there are also differences in the form of the skull, and some other bony structures, in the shape of the chin, and in the muscular and fatty development of different parts of the body. Nevertheless such differences are small indeed, when compared with those which exist between most species of apes and any other beasts whatever.

Similarly, the functions which the human body performs are similar to those performed by the body of the cat,† and, as regards the nervous system, we have,

---

* See *ante*, p. 217. † See *ante*, p. 221.

through observations on injured men, the best evidence of the existence in them of "reflex action." A man whose spinal cord has been greatly injured loses all power of both sensation and voluntary motion in those parts of the body supplied by nerves which come forth from the spinal cord below the place of injury. At the same time movements of those very parts (often exaggerated movements) may be produced by pinching, burning or tickling such parts, without the patient having any corresponding feelings, or any consciousness of the movements which he may see, that his own limbs are thus made to perform.

But while the nervous system of man ministers to all those feelings, single and associated, which result in such sense-perceptions and "consentience*" as the cat possesses, it also ministers to much higher powers than those which any beasts possess. We can describe a beast, but no beast can describe a man. The most important difference, then, between man and all the other objects of nature we have yet considered, is the difference which exists between his power of conscious thought, his *intellectuality*, and all the powers possessed by any other creature of which our senses can give us cognisance.

The study of this intellectual power and of the lower power of mere feeling, sense-perception, &c., which accompany it, constitutes a distinct study, which is usually designated *Psychology*. This study not only differs greatly from all other studies, as regards the object to which it is directed, but it differs absolutely from them as to the mode in which alone it can be carried on. In every other study our attention is

---

* See *ante*, p. 227.

directed to some external object or action, but in studying psychology we have to direct our attention inwards upon our own thoughts and feelings, and upon the actions of our own mind. Feelings and thoughts can, of course, be directly known only by the being who has them, and who knows that he possesses them, for without such knowledge he could not intentionally and deliberately examine them.

Now, on turning our mind inwards, we can, to begin with, recognise two very distinct kinds of mental activity. Let us suppose we are walking through a wood, while thinking about the study of nature. We may then recognise that a series of thoughts, of which we are conscious, has been passing (as the phrase goes) through our mind, while, at the same time, our feet must have received a series of sense-impressions from the ground walked over, to which sense-impressions we paid no attention at all, though, when we advert to them, we can recognise them as having existed. These two orders of mental activity, (1) *sensations* and (2) *thoughts*, are types of two distinct classes of mental activities—two faculties which we possess; the lower of these is our *sensitive activity* (a faculty which we share with other animals), the higher is our *intellectual faculty*, which, as far as we have been able to ascertain, is one possessed by no other animal whatever. This is probably the most fundamental and important of all the distinctions to be made in the study of the mind; for without it no accurate and satisfactory knowledge is possible of what the mind of man really is.

Now, in the first place, we have the power of recognising the existence of whatever we perceive to exist—to recognise that it really exists, to recognise its *being*. This idea, the idea of "being," is an idea which we must possess in order to be able to perform any intellectual act whatever. Most persons never think about it, and many readers may be surprised that they have had it all along without ever recognising it. But though not itself at first adverted to, it is by that idea alone that all other ideas are intelligible to us—as light, though itself unseen, makes all other things visible. If we cannot perceive that anything "is," we cannot, of course, perceive anything at all. But ordinarily, and especially in the beginning of our intellectual career, our attention is directed to real concrete external objects, in perceiving any one of which our minds acquire two distinct experiences: (1) the intellectual apprehension of the object perceived; and (2) the sensations, ordinarily unnoticed, which serve to make that object known to us. If the reader will consider for himself the action of his own mind, he will perceive such to be the case; thus, for example, should he, when reading this, have lately met a carriage with some friends of his in it, let him ask himself what was present to his mind at the time. He will say that the presence of the carriage and his friends was what he directly perceived. Of course, in order to perceive them, he must have experienced certain sensations: his eyes saw various patches of different colours, and his ears heard the sound produced by the wheels, the horse's hoofs, and his friends' voices. But he never adverted to these sensations at the time he felt them, though he can turn his mind back and recognise that they were then present to his sensitive faculty. His intellect was not occupied about his sensations when he perceived his friends, so that his sensations, though affecting his sensitive faculty, were not themselves perceived. Sensations are the *means*, not the *object* of perception. They hide themselves from our notice, in giving rise to the perception they elicit. They can only be recognised by an express turning back of the intellect upon them. Ordinarily they remain unnoticed, and we only perceive the thing they reveal—things they "represent," in the sense of *making them present* to the intellect. If we enter a library we do not see "images of books in rows" but the very books themselves. Perception is not inference from sense-impressions; for, in the first place, we do not attend to them, while if we are doubtful about any object, we "make sure" of it, not by any reasoning about sensations, but by merely tightening, as it were, our sensuous grasp of an object and focusing our sense-impressions more carefully. Intellectual perception, then, is a natural, spontaneous, and unconsciously made, interpretation of sensible signs, by a special power of our intelligence. Such an "interpretation" is an act of mental conception, and "concepts" are the simplest elements of our intellectual life. A *concept* contains implicitly a judgment. Thus, *e.g.*, the concept "bright sun" contains, implicitly, the explicit judgment "the sun is bright" and explicit judgment comes next after concepts. The most elementary complete acts of the mind are, then, explicit judgments, which are acts of the intellect, as it were, itself sitting in judgment on two concepts, and pronouncing as to some relation which exists between them — as, that they agree or disagree : in fact stating *explicitly*, what has been already *implicitly* seen.

Every object which we perceive, possesses a number of different qualities—shape, size, colour, hardness, &c., and acts on our sensibility accordingly. Now, our attention may be directed to various qualities, according to the different circumstances of each case, and then these qualities may be distinctly recognised as really being qualities of the object observed.

The power by which we thus ideally separate qualities is called *abstraction*, and by it our mind isolates (in order to apprehend them distinctly) the various qualities and conditions which really exist in any object perceived. Thus if the object perceived be a horse, we may notice that it has the characters of "a quadruped," that it is "a living creature," that it is "a solid body," or at least that it is "a something." Each such conception, though applicable to a multitude of individuals of the same kind, is a conception which, considered in itself, is *one*. It is a single notion, not of any one separate and subsisting thing, but it refers to a group of objects, to each of which the notion is applicable. It is an abstract idea of a lower or higher degree of abstraction; thus the term " horse" is an abstract idea formed through all we may have learned about horses. The term quadruped is more abstract, and is applicable to a vastly larger group of creatures. The same is again the case with the abstract ideas "living creature" and "solid body," while in the idea "something," which is the idea of "being," we arrive at the highest possible degree of abstraction, the most abstract of all "abstract ideas."

That feelings and "ideas" are fundamentally distinct,* is shown by the fact that the same idea may be called up in the mind by either sight, hearing, or feeling (*e.g.*, the idea "triangle"), while one set of sense impressions may give rise to a great number of different ideas, as the sight of a single photograph of the Queen may give rise to the

---

* The reader is referred for much further information on this subject to our work entitled "The Origin of Human Reason." Kegan Paul, Trench & Co., 1889, p. 45.

idea (1) of her Majesty herself; (2) of Royal rank; (3) of a woman; (4) of a human being; (5) of likeness; (6) of chemical action; (7) of the sun's actinic power;* (8) of the effect of light and shade; (9) of paper; (10) of an inanimate object; (11) of substance; and, finally, (12) of being or existence.

Feelings, again, can never reflect on feelings, but thoughts often reflect on thoughts. The vividness of a feeling (deafening sound or blinding light) may destroy the power of *sense*-perception but no vividness in an idea will mar *intellectual perception*. It is impossible for ideas to be too clear and distinct. Feelings become associated† according to the order in which they have been before felt, but ideas may become associated according to their intellectual affinity.

No efforts of our imagination, moreover, can ever exceed sensuous experience. We can never imagine what we have not felt in itself or in its elements; but it is quite otherwise with ideas. We have the idea "experience," but "experience" was never felt; and it is the same as regards an act of seeing, hearing, or of any other sense. The idea of "an act of seeing" is one thing, but the act of exercising a sensitive power is quite another. Nevertheless, though our thought can thus outrun our sensible experience (the experience we gain through our senses), it is impossible for us to entertain even the most abstract thought, except by the help of some mental image of things sensibly experienced—an imagination which serves as a support whereby we can mentally attain things beyond experience. The mental images which generally serve to aid us in our highest conceptions, are mental images of words spoken or written.

---

* See *ante*, p. 111. † See *ante* p. 227.

Our knowledge of our own mental and bodily activity, our *consciousness*, lies at the foundation of our whole intellectual life, as the parallel affection of our lower mental nature "consentience"* is at the foundation of all our sense perceptions. That we are conscious, is an ultimate fact of our being, our certain knowledge and perception of which no one can dispute. In so far as it is a fact—a state or quality of our being—it can, like other qualities, be mentally abstracted, and "consciousness" is thus both a fact and also an abstract idea gained from our own perception of our own self-knowledge.

Consciousness, though existing at each instant, is, in its very essence, continuous and conscious of its own persistence. We each of us know, and are conscious, not only that we are doing whatever we may be doing (as the reader is conscious that he is reading this page), but also that we began to do it and were doing something else before we so began. The supposition that consciousness could be composed of an aggregate of different "states" of consciousness is an absurdity. Such separate "states," if each were aware only of itself, could not constitute that consciousness which we know ourselves to possess, and which is aware of itself as continuous and successive. Consciousness is thus a persistent intelligence which, as from a fixed point, reviews the procession of events, and recognises them as severally belonging either to the order of ideas, or to what it recognises as real, external existences.

It is thus manifest that we have a certain power of directing our *attention* upon our feelings, or our thoughts, and of *reflecting* on them and comparing them, as well as of voluntarily attending to any ex-

---

* See *ante*, pp. 227 and 252.

ternal object—an act simulated by the merely sensitive activity of animals.*

Conscious reflection and attention also accompany and serve the next of our higher mental powers to which we desire to direct our readers' attention—namely, *intellectual memory*. To this the merely sensitive memory of a beast bears a certain analogy, which, however, is very remote, since it lacks the most essential character of the higher faculty, which is that it should be conscious. Evidently we cannot be said to "remember" anything unless we are conscious that the thing we so remember has been present to our mind on some previous occasion. An image might recur to our imagination a hundred times; but if at each recurrence it seemed to us something altogether new and unconnected with the past, we could not be said to remember it. It would rather be an example of extreme forgetfulness.

There is yet a further distinction between sensuous and intellectual memory. Every now and then we direct our attention to try and recollect something which we know we have for the moment forgotten, and which we instantly recognise when we have managed to recall it to our recollection. But besides this voluntary memory, we are sometimes startled by the flashing into consciousness of something we had forgotten, and which we were so far from trying to recollect, that we were, when it so flashed into consciousness, thinking of something entirely different. Mental movements of this kind may be distinguished, as *reminiscences*, from those which follow a voluntary search after things temporarily forgotten, which are *recollections*. It is obvious, however, that neither of these kinds of memory can exist without consciousness. Two

---

* See *ante*, p. 228.

other of our higher faculties are our two powers of *reasoning* or *inference*. One of these is commonly termed *induction*, and it is the process by which we are enabled to form judgments more or less probable, or even certain. Thus when we have examined many kinds of animals belonging to one group, *e.g.*, the group of opossums, and found that they all possess certain peculiar bones, we judge that if a new species is discovered it will also possess such bones. Such a judgment, however, can rarely be an absolutely certain one. But we may be certain in some cases. Thus by the study of many kinds of rock, we may recognise the truth, by means of fossils found in them, that different animals have inhabited the earth at different times. A further development of this kind of mental activity leads to a process which is most fruitful in promoting scientific progress—namely, the formation and verification of *hypotheses*. This process consists of guesses, based on a certain number of anterior observations, which guesses are subsequently tested by examining whether certain facts, specially selected as tests, confirm such guesses or not.

The other process of reasoning is termed *deduction* or *ratiocination*, and will be considered in the next chapter wherein we shall introduce the student to the elements of logic, and also make some additional remarks about induction.*

Amongst our many perceptions, one supremely important one is our perception whether or not anything is true—our perception of truth. The truth of every proposition of every science with which we have so far been concerned, consists in its conformity with fact. "Truth" is a conformity of thought with things, and

---

* See *post*, p. 309.

our power of apprehending such conformity is at the root of our whole intellectual life. The mental acts by which we perceive that anything is certainly "true," or has been found certain (like the acts by which we apprehend our own existence, or the existence or qualities of the objects which we directly perceive), are called *intuitions*. As they essentially pertain to our intellect, they are also called "intellectual intuitions."

Another of our perceptions of the highest practical importance is our perception whether any given action is, under its circumstances, a "good" action or not—our perceptions of *goodness*.

This perceptive faculty of ours must be carefully distinguished from keen feelings of sympathy, of shame, or of regret. The idea of "goodness" is quite distinct from the ideas of "utility" or "pleasure," and is a perception or "intuition" of *duty* as concerns ourselves or others. The radical distinctness which exists between our idea "goodness," and every other conception, is easily shown by an example. Let us suppose that anyone is told he "should pay his tailor," and the truth of the saying is disputed by him, the only possible way of trying to convince him as to his duty in that respect, would be to call his attention to some more general moral precept, such as "every man is bound to pay his debts." If this again were disputed, we might further urge: "A man is bound to satisfy obligations he has voluntarily incurred," and so on. In every step we take to explain why a duty should be performed, there must always be a further and more elementary declaration of duty, until we come to some assertion of the kind the truth of which is admitted as self-evident. This proves conclusively that no judgment as to moral obligation could ever be, or could ever have been, developed from mere likings or dislikings, or from pleasurable or painful feelings occasioned by the good-will or hostility of our fellow creatures.

The last of our faculties to which, in this elementary work, it seems to us indispensable to call attention, is our faculty of *will*. Our acts of will, our volitions, may be of two different kinds. They may either be (1) acts in which we simply follow, without any deliberate choice, our spontaneous inclinations; or (2) acts in which, after full deliberation, we elect to follow a course opposed to that towards which the balance of the attractions and repulsions acting on us would lead us to pursue. The common sense of mankind leads them to perceive this distinction, for when a man has lost this power of voluntarily " choosing," he is said to be *not accountable*. The conscience of every reasonable man assures him that he has, at least occasionally, a power of voluntarily fixing his attention upon one thing rather than another; and that he can, at any rate sometimes, act in oppositon to a strong temptation to violate duty. Our reason tells us that men are right in declaring that no moral blame can be attached to actions over which a man has no choice but which he is simply compelled to perform.

Thus over and above those powers which we possess, in common with the higher animals, we possess *self-consciousness* and powers of *intellectual perception, memory* and *reflection,* of forming *abstract ideas* (such as being, substance, cause, activity, passivity, self, not-self, difference and succession, extension, position, shape, size, number, motion, novelty, dubiousness, necessity, agreement, disagreement, &c), and of making abstract judgments. We also possess a power of *reasoning*, of *apprehending truth* and *goodness*, as such, and of *willing* in conformity with, or in opposition to, our moral perceptions. Further we possess a power of *language* entirely different from that of any other animal, as will appear at the end of this chapter.

As we are unable to entertain any thought without the accompaniment of some mental image or *phantasm* (as it is sometimes called), so we are unable to communicate our thoughts to others without giving them some audible or visible signs, and this fact makes language absolutely necessary for a social being such as man.

But man differs from other animals, in the long time he takes to attain the power of self-preservation and to reach maturity. Long absolutely dependent on parental aid, he requires the society of his fellow-creatures in order that he may attain anything like his normal intellectual development. It is from his fellows that he learns the use of language, without the possession of which it is impossible to make any considerable advance in knowledge.*

In the present chapter, it only remains for us to examine briefly what appears to be the intellectual nature of the lower races of mankind (with a view to forming a judgment as to whether they possess the same essential powers of mind as those which we ourselves enjoy) and to consider *language*.

Mankind at the present moment, consists of a great diversity of tribes and races, aggregated, partly into larger natural groups, and partly into political aggregations—states or nations. Each tribe, each race, each group of races, and each state or nation, has, of course, its separate history and its greater or less antiquity, its

---

* See "The Origin of Human Reason," pp. 166-171.

customs, sentiments, ideas, and language. But, as before said, they all agree in possessing a closely similar bodily structure.

All existing men supplement their natural bodily powers by the use of tools and weapons. The weapons of even very rude savages are commonly ornamented, and art in a rudimentary form may be said to be universally diffused. All tribes of mankind, without exception, possess the faculty of rational speech, and the language of gesture is very similar amongst savage tribes all over the world. A great deal has sometimes been made of the alleged inability of some savages to count more than five, or even three. The asserted facts are doubtful, but anyhow no one pretends that there are savages who cannot count as much as that. But to really count at all implies the possession of the idea of number, which is a highly abstract idea in itself and implies* much else.

For the idea "number" implies comparison, with a simultaneous recognition of both distinctness and similarity, although, of course, it is not necessary that the fact of possessing such apprehensions should be expressly adverted to. No two things could be known to be "two," unless we apprehended that they had "existence," and that while they were numerically distinct, they agreed in possessing a certain degree of similarity—that they were things which belong to the same order.

That many savages are frequent or habitual liars is an assertion which has been often made by travellers, while individuals, or exceptional tribes, have sometimes been praised for truthfulness. There can be no question,

---

* See "The Origin of Human Reason," pp. 81-91.

however, but that all men understand what stratagems, deceits, and lying are, and this they cannot have, without possessing a comprehension of "truth."

It is an unquestionable fact that different groups and sections of mankind differ as to their estimate of the moral character of certain actions; but this does not prove that there are any tribes of men who are devoid of all moral perceptions. The prevalence in any tribe of practices which shock us, does not prove that they have no apprehension of morality. Men are not necessarily devoid of it because they draw their ethical lines in different places from those we do. The most horrible actions (such, *e.g.*, as the deliberate slaying of aged parents) may really be the result of true moral judgments formed under peculiar conditions. It is said to be done by some savages in obedience to the wish of their fathers and mothers, who think thereby to escape further suffering in life, and to procure prolonged happiness after death. Their parricidal children reason accurately from mistaken principles—they make mistakes as to matters of fact. Men certainly do not always agree, even in England, about the application of moral principles: what they agree about is that there *are* moral principles. Thieving may be, here and there, encouraged and advocated, yet dishonesty is nowhere erected into a principle, but is reprobated in the very maxim "honour amongst thieves." Frightful cruelty towards prisoners was practised by the North American Indians, but it was towards *prisoners*, and cruelty was never advocated as an ideal to be always aimed at, so that remorse of conscience should be felt by any man who happened to have let slip a possible opportunity of inflicting torture.

Men have often thought it "right" to do things which were in fact unjust, but they have never thought that actions were "right" *because* they were "unjust," or "wrong" because they were "just."

One of the clearest of all ethical judgments is that as to "justice" and "injustice;" and, by common consent, the native Australians are admitted to be at about the lowest level of existing social development, while the Esquimaux cannot be said to be an elevated race. But Australians (at least some tribes of them) are of opinion that crimes may be compounded for by voluntary submission to punishment to be inflicted by those who have been injured. A criminal will thus submit himself to the ordeal of having spears thrown at him or thrust through certain parts, such as the calf of the leg or under the arm, by those he has injured. But if the punishment exceeds the extent sanctioned by the native code, or is inflicted in a wrong place, then the man so offending becomes liable to punishment in turn.

A yet stranger example of the existence of distinct moral perception amongst very rude people is furnished us by the Greenlanders. Should a seal escape with a hunter's javelin in it, and be killed by another Greenlander afterwards, it belongs to the former. But if, after the seal is struck with the harpoon and bladder, the string breaks, the hunter loses his right. If a man finds a seal dead with a harpoon in it, he keeps the seal but returns the harpoon. Any man who finds a piece of driftwood can appropriate it by placing a stone on it, as a sign that some one has taken possession of it.

The inhabitants of Tierra del Fuego are, if possible, more wretched savages than the Australians. Yet even these have an unmistakable perception that to waste human food is wrong, and may justly incur retribution from a higher power. Although there may be savages, as there may be Englishmen, who seem devoid of moral feelings or ideas, there is no evidence of the absence of moral perceptions in any tribe of human beings. A power of occasionally distinguishing an element of right and of wrong in certain actions is a power possessed by all normally constituted human beings. Even the worst men can recognise such a character in acts of treachery done *to them* by their own special associates.

Man also seems habitually to have formed some judgments about matters commonly called "religious." That is to say, all races of men have the idea of the existence of some kind of personal relation between themselves and some invisible being or beings—malevolent or benevolent—with more than human power and possessing faculties more or less analogous to human intelligence and will. The relations which exist, and should exist, between human beings constitute the science of *sociology*. The most general expression, then, for natural religion is, a code regulating the relations which should exist between human beings and invisible, non-human intelligences. It is a sort of "supernatural sociology;" it merits that term the more, inasmuch as some regulations as to the conduct of men amongst themselves generally enter into it. The universal tendency of even the most degraded tribes to practices which clearly show their belief in preternatural agencies is too notorious to admit of serious discussion, while the probably all but universal practice of some kind of funeral ceremony speaks plainly of as widespread a notion that the dead in some sense yet live. Many funeral arrangements are evidently intended to impede, or render impossible, the return of the deceased to the home he has left.

It is not, of course, meant to affirm that all savage men, any more than all civilised men, believe in a future life, or in one or many gods; but nevertheless, the reality or probability of some quasi-social relations between men and invisible intelligences is a common attribute of mankind.

Considering all the known facts, it is, in our judgment, certain that all men possess an intellectual, as well as a bodily, nature which is essentially one, however either may vary in minor details. As has been said by an authority in these matters: * "There are multitudes of negroes who reach as high an intellectual level as many Europeans. Man's craving to learn the causes at work in each event he witnesses, the reason why each state of things he surveys is such as it is and no other, is no product of high civilisation, but is a characteristic of the human race down to its lowest stage. Among rude savages it is still an intellectual appetite, the satisfaction of which absorbs many a moment not engaged in war, sport, food, or sleep." What can more plainly indicate the presence of this intellect than the apprehension of those very abstract ideas indicated by questions as to "what," "how," and "why." The investigation of such questions constitutes, as we shall see, the highest form of science, and depends upon man's power apprehending the ideas "causation" and "a cause," ideas which are born of our knowledge of our own powers of mental determination, and of giving effect to such mental determination, by our external acts, as also by the resistances our efforts meet with.

The study of man, of our own nature, especially as revealed to us by psychology (that is, through the interrogation of our own consciousness), reveals to us the most wonderful and important novelty wherewith our elementary study of the universe has made us acquainted. In

---

* Mr. Tylor.

the preceding chapter we first met with two activities which were wonderfully different from any to be recognised as existing in the inorganic world, the first of these was life and the possession by certain beings of a power of internal growth ultimately resulting in the generation of new individuals. The second wonder was that faculty of feeling, which seemed to be bound up with the possession by an organism of a certain kind of tissue, so that only a section of the whole organic world could, with certainty be deemed to possess it. But the higher animals, we saw, have such complex and definite sensitivity that it is evident they not only possess special senses, but can sensuously perceive objects about them, that they possess imaginations and a kind of memory, with emotions and anticipations of feelings yet to be experienced.

In the present chapter we make an advance upon what we have previously studied, which is far greater than the advance we made in passing from the non-living world to the world instinct with life and feeling. For now we have entered upon the world of intellectuality—the world of thought.

And here we put aside all questions of origin and essential nature, as we before did with respect to life and the physical forces. When treating of heat, light, &c., we declared that an elementary work like the present was not the place wherein to consider the problems of what such physical forces in themselves might be*, and in describing life† we forbore to enter upon the question what life is, as similarly beyond our present aim. So as regards intellect and thought, it matters not to us, here and now, how or whence they came to be, how early in life they may show themselves, or how

---

* See *ante*, p. 86. † See *ante*, p. 189.

universally they may exist amongst our fellow men. If only we know that we ourselves are conscious and can reflect, mentally abstract, and judge, and know that we are so doing, that is enough to make us absolutely certain we possess a power, different indeed from each and all the powers which have been previously spoken of in this little work. Moreover, this knowledge is the most absolutely certain of all knowledge. However we may doubt, we cannot doubt that we think while we are thinking, whatever may be the object of our thoughts. And thought is our ultimate test of truth. Even in investigating the properties of material bodies, it is to thought we must ultimately appeal. Only by that can we know what we may have done, and it is that which judges between conflicting indications of different sense-impressions. Our senses are truly tests of certainty, but not *the* test. Self-conscious, reflective thought is and must be our last and ultimate criterion.

And now let the reader consider what his own intellect tells him about its own powers and activities as revealed to him directly in his consciousness. He will recognise that his intellect (that is, he himself) exists continuously, so that he knows that it is that same intellect of his which both began to think of this question, and now still continues to think about it. He will also recognise that he knows objects and persons about him, and what is happening to them before his eyes, while he all the time remains something distinct from them. He can reflect over the different things he remembers to have thought of during the morning, and recognise that they have constituted a series of thoughts. He can thus think of them as a whole—a series—or he can select one of his thoughts, or a group of them, for reconsideration. He knows also what he is about, what he is doing, or what may be being done to him. As to external objects and events, he can compare them and see some of the relations in which they may stand to one another, following up different lines of thought as he will.

But such a power as this intellect of his, aware of all these things and mentally present to them all, cannot be made up of parts, but must be the most perfect and simple unity we can conceive of. It cannot therefore be either a material substance or a physical force, and still less any combination of such substances and forces, to which it presents the greatest contrast. Nevertheless we each of us have also the structure, the faculties and the feelings of a sort of ape. Therefore each of us is a unity with two sets of faculties; (1) one essentially sensuous and material, the source of all bodily functions and animal feelings; the other (2) essentially intellectual and immaterial, which can examine and judge the world about it and also its own activities.

LANGUAGE.—Next we must consider that most fundamental and distinctive character of man as an intellectual being—namely, his power of speech and of making known his meaning by intellectual gestures.

We have noted* how the higher animals by vocal cries and bodily movements, or gestures, can give expression to the various different feelings by which they are animated; as also that human language is something entirely different† therefrom. The time has now arrived for paying direct attention to this very important subject. That it is very important is manifest, for it is impossible for us to make known our thoughts to others, save by the help of bodily signs, vocal or other. But it is more important

---

* See *ante*, p 229. † See *ante*, p. 263.

still, for, as before said,* we cannot even think without some imagination of things noted by the senses, and especially of words, as spoken, heard or read, or of gestures as seen in reality, or in pictures, or as felt. We almost always think in words, though we may think by the aid of objects or actions pictured by the imagination.

We have to consider language, in the ordinary sense of that term, as a medium for expressing ideas and intentions, asking questions, stating facts, and carrying on conversation. Since then by "language" we ordinarily mean spoken articulate sounds, serving for intellectual intercourse, we will begin by examining some very simple expressions of the kind. Let us suppose two men to be standing under an oak tree, and that this tree begins suddenly to show signs of falling, they will fly from the danger, and they may utter cries of alarm, and by their cries and gestures they may give rise to sympathetic feelings of alarm in persons who happen to be near the spot. In so far as they do no more than this, their language (whether vocal or of gesture) is but of the same kind as the language of emotion in the lower animals.

They may, however, cry out "That oak is falling," that is, they may give utterance to intellectual language; for those four words express and embody much more than any mere feelings or emotions. They express and signify abstract ideas.

(1) The word *oak* is a conventional sign for the idea, or conception, of that kind of tree, and is an abstract term applicable to every actual or possible oak.

(2) The word *that* is one which serves to separate off,

---

* See *ante*, p. 257.

in the mind, the one particular falling oak from all others. It implies the idea of a unity of a different sort from the unity implied by the word "oak."

(3) The word *is* denotes the most abstract of all abstract terms, the idea of "existence" or "being," the significance of which has been before pointed out.*

(4) The word *falling* is a term denoting another abstraction—the conception of a "quality" or "state." The idea is one which is evidently capable of a very wide application, namely, to everything which may fall. Yet the idea itself is one single idea.

A word by itself has but a very imperfect signification. It needs the addition, expressed or implied, of others to give it full meaning—to explicitly † express a judgment. Such a congeries of words is *a sentence,* and the four words "That oak is falling," is one such significant set of words.

What is true of this single sentence is true of all sentences. All human language (apart from mere emotional manifestations) necessarily implies and gives expression to a number of abstract ideas. It is impossible for even the most brutal savage to speak the simplest sentence without having first formed for himself highly abstract ideas. Wherever, therefore, language exists, there also must exist the power and the active exercise of mental abstraction.

All our words express abstract ideas, except proper names, and words denoting individuals, such as the words "I" "thou" "they," &c. Words which denote individual objects, real or ideal, are termed, as we all know, "substantives," as a *parrot,* a *desire.* A word which indicates that anything acts or is acted on, is, as

---

* See *ante,* p. 253.   † See *ante,* p. 255.

the reader is, of course, well aware, called a "verb," as, the parrot *screams*, or, he *shot* the parrot. Of course nothing can act or be acted on, unless it exists, but there is a special verb, called the substantive verb, which expresses existence. It is the verb *to be*, and, in its form "is," it asserts what has above been pointed out. When it asserts a relation between one thing and another, it is spoken of as the *copula*. Thus in the sentences, "That man *is* a father," "The sky *is* blue," and "That oak *is* falling," the term "is" merits that appellation.

Terms which denote qualities or states of objects are "adjectives," or verbs in that form which is called a "participle"—as in the last two sentences given as examples, namely, "The sky is *blue*," and "That oak is *falling*."

Elementary as this work is intended to be, it is not our object to teach in it such things as orthography or grammar, we will therefore pass on to further point out wherein the essentials of human language consist.

The faculty of abstraction must, as before said, be possessed by every one who speaks. But that faculty is also possessed by men who do not speak. Various kinds and degrees of dumbness may arise from different forms of defective memory as to words, due to different physical defects of brain-structure, such defects impairing those powers of feeling and imagination, on the integrity of which the exercise of our intellectual faculties depends. The absence of words does not necessarily imply the absence of ideas. Very wonderful are the gestures of deaf-mutes[*] which make it unquestionable that human beings may possess distinct intellectual conceptions in the entire absence of spoken

---

[*] See "The Origin of Human Reason," pp. 138–171.

words. Abstract ideas, then, can exist without such words, but there is no evidence that they can continue to exist without some embodiment, some form of language, some corporeal expression, either by voice or by gesture. Language therefore is a *consequence* of thought, and abstract ideas are indispensable preliminaries to language. We see this in our common experience. When in the cultivation of any science or art, newly observed facts, or newly devised processes, give rise to new conceptions, new terms are invented to give expression to such conceptions. Thus new words arise as a *consequent** and not as an *antecedent*, of such intellectual actions. New terms are always fitted to fresh ideas, and not fresh ideas to new terms. That language is dependent on thought, not thought on language, is demonstrated for us by the lightning-like rapidity—a rapidity far too great for words—with which our minds may detect a fallacy in an argument. This instantaneousness is not the mere mental ejaculation of the word "no," but is due to our having apprehended the relations existing between different portions of the argument. The most rapid cry or gesture of negation, is often the sign of intellectual perceptions which would require more than one sentence fully to express, but which are perceived too rapidly for even the *mental* repetition of the words of such sentences. Nevertheless, these intellectual perceptions show themselves by bodily signs—sounds or gestures—and even all our silent thought is carried on by the aid of some imagined bodily signs, without which, as we before observed, we cannot think. Human language seems

---

* See a correspondence with Professor Max-Müller, cited in "Origin of Human Reason," pp. 99–117.

quite unable to grow, or even to endure, without some embodiment, without corporeal expressions of some kind. Thus language of word or gesture is the necessary means of human thought as well as its necessary consequence.

The mental and bodily sides of language are so intimately united that, though the mental side is anterior, it at once seeks, as it were, to incarnate itself, and, under normal circumstances, does incarnate itself, in corporeal expression. Deaf mutes will spontaneously evolve a gesture language through which they can understand each other and communicate their ideas. Rational conceptions, therefore, can evidently exist without words, but rational words cannot exist without conceptions or abstract ideas. But though a language of gesture may be elaborated, and even carried to much perfection, as we see in certain ballets, yet spoken language is so enormously superior to that of gesture that no comparison can be made between the two vehicles for the expression of human thought.

The intellect is the common root from which both thought and language (whether of speech or gesture) spring, and thenceforth continue and develop in unseparable union.

Language, then, is of two radically distinct kinds—(1) the language of feeling which we possess in common with animals; and (2) the language of the intellect which is absolutely peculiar to man.

There are also various subdivisions of each of these two kinds of language.

Of the mere language of the emotions and of feeling we may have:

(1) Sounds which are neither articulate nor rational, such as cries of pain or the murmur of a mother to her infant.

(2) Sounds which are articulate but not rational, such as many oaths and exclamations, and the words of certain idiots, who will repeat, without comprehending, every phrase they hear. This habit is parallel to that power of irrationally emitting articulate sounds which some birds possess.

(3) Gestures which do not answer to rational conceptions, but are the bodily signs of pain or pleasure, of passion or emotion.

The subdivisions of the language of the intellect are:

(1) Sounds which are rational but not articulate, such as inarticulate ejaculations by which we sometimes express assent to, or dissent from, given propositions.

(2) Sounds which are both rational and articulate, constituting true "speech."

(3) Gestures which give expression to rational conceptions, and are therefore corporeal, but not "oral," manifestations of abstract thought. Such are many of the gestures of deaf-mutes, who, being incapable of articulating words, have invented or acquired a true gesture-language.

(4) A peculiar modification of gesture which takes the shape of (a) pictorial signs of objects or actions; or (b) signs of sounds which, taken together, form words signifying abstract ideas—(i.) picture; or (ii.) written language.

Thus the essence of true, or intellectual, language is mental, and is a form of intellectual activity which, as the mental constituent of speech, may be distinguished as the "mental word" or *verbum mentale*. The other element of speech, that which gives it external expression, may, on the other hand, be distingushed as the "spoken word" or *verbum oris*. The latter ever follows the former, as is evident from that process which goes on in every science as it progresses, the process of inventing (as before said) fresh terms in order to denote new or more complete and better defined conceptions.

The mental word is normally in excess of the spoken word, as is shown by our use of *metaphors*. Had not the intellect the power of apprehending, through the senses, what is beyond the power of feeling, metaphor would not exist. It exists because speech is too narrow for thought, and because words are often too few to convey the ideas of the mind.

The mental word is also so rapid, it often happens that speech cannot keep pace with it. That such is the case as regards writing most readers have probably already noted.

But there is the closest connection between the mental and the spoken word, which ordinarily accompany each other most closely, as the concavities and convexities of the opposite sides of an undulating line. They are nevertheless sometimes abnormally separated through some physical defect, the tongue repeating other words than those the mind desires to give expression to. A paralysed man may possess the mental word though hindered from manifesting it externally by spoken words, or even by gestures. But normally, as just observed, the external and internal powers are inseparable. When the intellectual activity exists, it seeks external expressions of symbols—verbal, manual, or what not—by the voice or by gesture-language. Some form of symbolic expression is, therefore, the necessary consequence in man of the possession of reason, while it is impossible that true speech can for a moment exist without the co-existence with it of that intellectual activity of which it is the outward expression.

But language reacts upon the mind of him who uses it. Besides being an interpreter of thought, it exercises a powerful influence on the process of thinking. (1) It helps us to analyse new complex impressions by the distinctness and definiteness which it has given to our conceptions of previous impressions. (2) By memory or writing it enables us to preserve the results of mental activity for future service. (3) It greatly aids mental processes by substituting a short word for a very complex idea accompanied by a variety of mental images. Signs, when their meaning is definitely established, can be used as counters, and we can temporarily neglect what they signify while we are working with them till we come to a final result, as in the case of arithmetical and algebraic signs, as we saw in the second chapter.* (4) Finally language serves, as every one knows, as a means of communication, and so has become an enormous factor in mental development, since, as we have seen, the mental development of each man depends directly on his intellectual environment which appears absolutely indispensable to it, although neglected children and deaf mutes are said† to be able to form a sort of language for themselves.

Just as all races of men are men, and exhibit no great or essential structural or functional differences, so they all possess the power of speech, although that power is expressed in many different languages, each of which has its own method of expression. In Latin one word, *amavi*,

---

* See *ante*, pp. 12 and 23.
† See "Origin of Human Reason," p. 232.

serves to express what in English requires the three words "I have loved."

How the various sounds which in different languages express different ideas are agglutinated, isolated, and changed in form or "inflected," is a matter which cannot be gone into here. For such further information the reader is referred to the various works on the science of language, or *Philology*.

But objections to the essential unity of language have been made because certain differences exist in the mode of giving expression to the same ideas. Thus, it has been affirmed that some languages are defective as regards the substantive verb, and that therefore those who speak it have not the idea of existence. But if there are tribes of men who cannot say, "He is sitting," "She is standing," "It is falling," it is quite enough if they can say, "He sits," or "Stands she" or "It falls." Such expressions show that they have the idea. It is a man's *meaning*, and his power of making that meaning *evident*, which is alone important, the system of signs by which he expresses it is a relatively trivial matter.

It is not a multitude of words which constitutes the perfection of language. Some exceptionally endowed minds can, with a few pregnant words, bring to the minds of others, perceptions which could be conveyed by inferior natures only by means of long and laboured discourses. Therefore, the minimum of language which can co-exist with the due expression of thought is the best.

The ultimate facts as to language may be thus expressed:

(1) Thought is the root of every sort of intellectual language, and is anterior to outward expression of whatever kind.

(2) Language is the external expression of the *verbum mentale*.

(3) The simplest elements of thought being, as we have seen,* an "implicit judgment" or "concept," the simplest element of language must be the external expression of an implicit judgment—*i.e.*, a *word* or *term*.

(4) The most elementary complete act of the mind being a judgment,* the most elementary complete expression in language must be a judgment expressed in words or other signs—*i.e.*, an *enunciation* or *proposition*.

This brief sketch of the most important characteristics of man, as distinguished from other animals, must suffice as an introduction to the elements of the science of human nature, or *Anthropology*. Without such elementary knowledge as we have here attempted to convey, any study of nature, including man, must be futile and deceptive. Having now indicated what are the essential characters of the human intellect, we must devote the following chapter to an exposition of the main laws which govern the employment of our wonderful faculty of reason.

---

* See *ante*, p. 255.

CHAPTER VIII

LOGIC

Logic is the science which treats of the laws that regulate human thought. It is the science which elucidates the mode in which thought must be carried on in order that we should arrive at truth. Nevertheless it does not concern itself with the truth of the thoughts themselves —their conformity with external things—but only with the mode in which thought must be employed in order to attain two ends. The first of these (*a*) is the avoidance of certain errors which would necessarily arise were the rules of logic violated. The second (*b*) is the manifestation of truths which are involved in, and depend upon, the recognition of other antecedent truths, from the truth of which they necessarily follow as consequences.

The importance of logic as a key to all reasoning, and therefore to a large part of every science, is enormous. On this account, though we must refer our readers to special treatises for all but the elements of it, we propose to treat these logical elements at some little length.

The first *meaning*, the first *intention*, we have in thinking about, or naming anything, relates directly to whatever we so think about or name. Thus, for example, if we think (1) "a horse," or (2) "that horse is lame," or (3) "its lameness shows that the man who sold it cheated," in each of these cases our meaning or "intention" respectively is (1) the kind of animal we have formed a concept of; (2) the lameness we judge about; or (3) the dishonest action we infer to have taken place. But these three thoughts have also the quality of being thoughts of a special kind—belonging to one or other sets, or kinds, or groups, or *forms* of thought. Moreover, each of them belongs to a *different* group or form— they are different "forms" of thought. Thus the first (1) is a concept,* the second (2) is a judgment,* and the last (3) is an inference—a conclusion arrived at by a process of reasoning.

Thus each thought signifies, not only what is our meaning, or first intention in using it, but also, what "order" of thought it is. This latter signification is said to be the second meaning or second "intention" of each thought. Therefore logic is called the science of *second intentions*. By this it is meant that logic treats of the forms of thought, regardless of the matter to which such thought may refer. Thus, *e.g.*, in the thought "that horse is lame," the *form* of the thought is a "judgment." The *matter* of the thought is the fact of lameness, with which fact logic has nothing whatever to do. As therefore it has only to do with the forms which thought may assume, it is justly called "*the science of the formal law of thought.*"

But though logic has nothing to do with the material truth of thoughts, it has to do, as before said, with so conducting processess of thought as to secure that they shall not themselves be the occasion of error, but, on the contrary, aid in making explicit, truths which such thoughts may implicitly contain. Logic, therefore, as a guide to just and valid reasoning, is practical, and

---

* See *ante*, p. 255.

consequently is an *art* as well as a science—it is *the art of correct thinking and of valid inference.*

The science of logic, as "the science of the forms of thought," is divisible into three parts, each of which deals with one class of such forms. The first part deals with *conceptions*, the second with *judgments*, and the third with *reasoning*.

The first part which thus deals with "conceptions," is also said to be occupied about *names*, or *terms*. This is said on account of the intimate association which exists between each thought, or *verbum mentale*, and the word which signifies it, or *verbum oris*.* The connection is so close that we may henceforth, in this chapter, deal with the spoken and written signs of thoughts as if they were the very thoughts themselves; although it should nevertheless be remembered that we are not treating of names *as such*, but as symbols of the thoughts they represent.

A "name," or "term," is a human, conventional articulate sound, used with an intention of signifying something. There are various kinds of names. They may be (1) *universal*—applicable to each and all of a class of objects, *e.g.*, "man," "true," &c.; (2) *collective*—applicable to a whole but not to its component members, *e.g.*, "army"; (3) *particular*—applicable to but a portion of some whole, *e.g.*, "some Indians"; (4) *singular*—applicable to one only, *e.g.*, "Julius Cæsar"; (5) *universal*—having the same signification when applied to different objects, *e.g.*, "green," as applied to foliage and a lady's dress; (6) *equivocal*—one sound having more than one signification. *e.g.*, "box"; (7) *analogous*—when one signification applies in an unequal degree to different objects. Such

---

* See *ante*, p. 277.

analogy may be (*a*) one of *proportion*, or (*b*) one of *attribution*. In the former case there is a certain agreement in the effects produced by quite dissimilar causes, *e.g.*, "the *staff* of life," the "*foot* of a mountain," the "*head* of an army," or, as we may say of a vessel, "she *walks* the waters like a thing of life." Words which have an analogy of *attribution* are such as denote a quality which primarily and properly belongs to one object but is attributed to another as bearing some relation (as of cause or effect) thereto, *e.g.*, *healthy* man, *healthy* exercise, *healthy* appetite, *healthy* climate; (8) *abstract* *—denoting some form or quality, considered apart from whatever object or objects it may pertain to, *e.g.*, "whiteness," "existence," "extension," &c.; (9) *concrete*—applicable only to things actually existing, or having existed, *e.g.*, "Descartes," "these sheep," &c.; (10) *absolute*—that is, a name the signification of which refers only to the object named, *e.g.*, "justice," "gold," &c.; (11) *connotative*—that is, necessarily implying something else, *e.g.*, "white," "rapid," "pianist," "son," &c. Terms which have a natural relation to one another, such as, *e.g.*, father and son, are spoken of as *correlative*. (12) *positive*, *e.g.*, "strong;" (13) *negative*, *e.g.*, "insentient," as applied to a stone; (14) *privative*, *e.g.*, "insentient," as applied to a man who has lost the power of sensation; (15) *transcendental*—names denoting ideas of attributes which are incapable of (or transcend) classification, because they are present in all cases and pertain to everything which there is to classify, *e.g.*, "existence."

Universals are words denoting classes of things, or attributes, or qualities, which may exist in many things. They relate to things as they exist, apart from

---

* See *ante*, p. 256.

all apprehension of them. A horse, and our idea of, and name for, a horse, remain the same, whatever may be the direction of our mind in considering such things, and the same is the case with regard to the quality "quadrupedal." As we have seen,* living things are arranged by science in a definite classification. Therein the creatures which constitute a genus are always the same, and the terms "genus" and "species" can never change places. But there are, in logic, certain things and names, which, instead of being constant, change according to the direction in which the mind is turned as it regards them, and their mode of classification changes correspondingly, and in this way classificatory groups may change in position and value. Thus we may think of "Jews," as a group. Then "Jews of Middlesex," "those of Essex," &c. &c., will be minor groups subordinate to the larger and dominant one "Jews." But we may also first think of "Jews of Middlesex," and then of those which are "reformed," and those which are "unreformed," which will then constitute subordinate groups, while the group "Jews of Middlesex" will change from being, as before, subordinate, and become the larger and dominant one. Similarly, the group "Jews" will change from a dominant group into a subordinate one, if we think of the larger group, "European." Now these conceptions of dominant and subordinate groups are respectively characterised by certain *differences*. Thus the difference between "Jews" generally and "those of Middlesex," consists in the latter inhabiting that county.

There are also certain qualities which are essential, or *proper*, to objects, and others which belong to them only

---

* See *ante*, p. 190.

accidentally. Thus a Jew is a man who believes himself bound to keep a Sabbath, and this is "essential," but he may be lame or have had his hair cut short, characters which are but "accidental" ones. So it is that, in logic, there are certain names which can only be asserted, or *predicated*, of things, according to the mode in which the mind regards them. They are therefore called *predicables*, and there are five of them—(1) genus; (2) difference; (3) species; (4) property; and (5) accident.

The predicable "genus" is applied to any larger group which includes other groups within it, to every one of which it applies, and each one of which is, in logic, termed a "species." Thus in the first of the above examples "Jew" is a genus which includes the minor groups or *species*—"certain men of Middlesex," "certain men of Essex," &c.—to each of which, that which the name of the genus denotes, is applicable. The term "Jews of Middlesex," again, is also a genus, including within it subordinate groups or species, such as "reformed Jews" and "unreformed Jews," while "European" may also be a genus, whereof Jew, Englishman, Russian, &c., are species.

Thus genus and species in logic are very different from genus and species as these terms are used by zoologists and botanists.*

In those sciences one set of groups is always genera, and others are always species, and none can be *both* genus and species. But in logic, groups are not always one or the other, but may be either, according to the direction taken by thought, as we see "Jew" may be a *genus*, with respect to "men of Middlesex," "Essex,"

---

* See *ante*, p. 190.

&c., but it may be equally a *species*, with respect to the concept and name "European."

Thus let the circle A (Fig. 54) represent the genus "Jew," and B the differentiating quality "man of Middlesex," then these by their intersection will constitute C—*i.e.*, the species "Jew of Middlesex."

On the other hand, A may represent the genus "man of Middlesex," and B the quality "Jew," with the same result with respect to C. Thus the genus and the differential quality or "difference" may change places.

Again B may be the genus "man," and A the quality

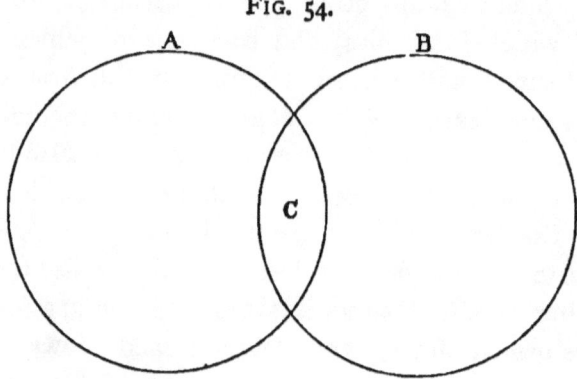

FIG. 54.

"hereditary observer of the Mosaic Law," when C will be "Jew" as a species.

A logical definition of any concept is formed by uniting the proximate, or lowest, genus in which it is contained, with its differential quality.

The lowest species can never be a genus, and consists only of the individuals which compose it. Similarly the highest genus can never be a species because, being the highest, it cannot be included, as a subordinate group, within any higher one.

Between the highest genus and the lowest species are a multitude of genera which are called *subalternate*, because (as in the above examples) they may be either genera or species according to the way in which they are used, owing to the direction taken by thought. Thus the term "corporeal" is a *species* of the higher group, the genus "substance," while it is a *genus* of the lower group, "living creatures," which is here a species. That species, however, becomes, in turn, a genus if we take it in connection with "animals" and "plants," both of which are species of the genus "living creature." "Plant" in turn will become a genus, whereof the species "flowering" and "non-flowering" plants are species, and so on.

The attributes of the higher group, can, of course, be affirmed of every member of all the lower groups which are included within it, as "corporeity" can be affirmed of all living creatures, all animals and all plants, and all the various more and more subordinate groups of either. But the higher the group, the fewer the number of attributes which can be affirmed of all its component members, although the larger it is, the greater the number of such component members will be. On the other hand, the smaller and more subordinate the group, the greater the number of common characters possessed by its component members. Thus the group "apple tree," though it comprises almost infinitely fewer component members than the group "living creatures," nevertheless possesses an almost infinitely greater number of common characters than the group "living creatures" possesses; for a vastly greater number of properties are common to all apple trees than are common to the enormous group "living creatures," which includes everything between a mushroom and a man. A difference of this kind is expressed by the terms *extension* and *intension*. "Extension" refers to the number of groups contained within a concept; "intension" to the number of common characters which any group we think of may possess. Thus the concept "living creature," has enormous "extension" and but little "intension," while the concept "apple tree" has very great "intension" and but little "extension"—since it includes no living creatures whatever save apple trees. Thus the greater the "extension" the less the "intension," and *vice versâ*.

A concept is defined in logic by means of the lowest genus which contains it, and the difference which separates it off from the other species of that genus, as illustrated in Fig. 54.

For all other matters which concern the first part of logic, the reader is referred to works upon that science.

The second part of logic treats, as before said, of judgments. What a judgment in itself is, we have already seen.* When expressed in words it is termed an "enunciation" or "proposition."

A *proposition* is an expression unambiguously affirming, or denying, one thing of another, and it consists of three parts, two of these are termed respectively (1) the subject; and (2) the predicate; the third is the copula.

The *subject* is that term of which the other is affirmed, as, *e.g*, in the proposition "virtue is admirable," "virtue" is the subject. The *predicate* is that term which is affirmed of the other—as "admirable" is affirmed of virtue. The *copula* is the term which connects, or couples, the subject and the predicate, and it is expressed by the term "is."

Each term may consist of many words, as, *e.g*, "To rise early in the morning and take a cold bath is healthy." Here, then, the eleven words which precede "is" together constitute one term.

---

* See *ante*, p. 255.

Propositions may be *categorical* or *hypothetical*, and the latter may be either *conditional* (as, if X then Y), or *disjunctive* (as, either X or Y).

"Categorical" propositions may be *universal* or *particular*, and each of them may be either *affirmative* or *negative*. Thus every categorical proposition is either:

A universal affirmative which may be represented by .... A.
(*e.g.*, all whales are mammals.)
A universal negative................. E.
(*e.g.*, no whales are fishes.)
A particular affirmative .... I.
(*e.g.*, some whales are toothed.)
A particular negative .... O.
(*e.g.* some whales are not toothed.)

Propositions which differ as to being universal or particular, are said to differ in *quantity*. Those which differ as to being affirmative or negative, are said to differ in *quality*.

Four kinds of opposition may thus exist between judgments, and they may be: (1) *Contradictories;* (2) *Contraries;* (3) *Subcontraries;* or (4) *Subalterns*.

```
      A      Contraries      E


  Subalterns    Contra  dictories   Subalterns
                Contra  dictories


      I      Subcontraries   O
```

When a universal affirmative and a particular negative (A and O) or a particular affirmative and a universal negative (I and E) are opposed, they are "contradictories," because they differ both in quantity (one being universal and the other particular), and also in quality (one being affirmative and the other negative): *e.g.*, "All men are reasonable beings" and "some men are not reasonable beings," or "Some men are able to rise in the air" and "no men are able to rise in the air." With *contradictories*, both cannot be true and both false, but always one true and the other false.

When a universal affirmative and a universal negative (A and E) are opposed they are termed "contraries." They differ from each other in quality only and not in quantity, because they are both universals; as *e.g.*, "All plants are sensitive" and "no plants are sensitive." *Contraries* may both be false, but they cannot both be true.

When a particular affirmative is opposed to a particular negative (I and O) they are called "subcontraries" and differ as to quality only, both being particular, and therefore the same in quantity; as *e.g.*, "Some men are soldiers" and "some men are not soldiers." *Subcontraries* may both be true, but they cannot both be false.

When a universal affirmative and a particular affirmative (A and I), or a universal negative and a particular negative (E and O) are opposed, they are named "subalterns." They are opposed in quantity only (each pair consisting of one universal and one particular proposition) and not in quality (one pair being both affirmative and the other pair both negative); as *e.g.*, "All men are mortal" and "some men are coal-heavers," or "No men have tails" and "some men have not snub-noses."

Subalterns may both be false and both true, or one may be false and the other true in either case.

A term is said to be *distributed*, when it includes every individual of the class to which it refers. Thus, when we say " fixed stars twinkle " and " no man can fly, " we mean to include the whole of the fixed stars in one case, and all men in the other.

In all universal propositions, the subject is distributed—as in the instances just given. In all negative propositions, the predicate is distributed—as " Men are not immortal " and " some whales are not toothed." " Immortal " is distributed because it refers to all men, and "not toothed" is distributed because it refers to all those whales which are referred to and included in the above proposition.

Therefore in universal negative propositions, both terms are distributed, while in particular affirmative propositions, neither term is distributed.

Propositions can be inverted; that is, the two terms of a proposition can be transposed, and this process is termed *conversion*. In order that this may be carried out validly, no term must be distributed in the converted proposition, which was not distributed before its conversion.

Since in universal negatives (E) both terms are distributed, *e.g.*, " No horses are winged creatures," and in particular affirmatives (I) neither term is distributed, *e.g.*, " Some horses are lame," therefore in both these cases the terms can be simply transposed and yet the propositions must remain true—as "winged creatures are not horses" and "lame are some horses."

A universal affirmative (A) may be converted by changing it into a particular affirmative (I), as, " All men are animals" may be converted thus: Some animals are men.

A particular negative proposition (O) can only be converted by a threefold process which consists in (1) changing the copula from negative to affirmative; (2) inverting the terms; and (3) changing the subject into the term contradictory to it. Thus, *e.g.*, the proposition:

"Some animals are not apes," may be changed by—

(1) Change of copula, "Some animals are apes."
(2) Inversion of terms, "Apes are some animals."
(3) Changing subject into its contradictory, } "Not apes are some animals."

The *third part of logic* concerns reasoning, which may be either *inductive or deductive*. Inductive reasoning has been and will again be referred to.*

"Deductive reasoning," or ratiocination, consists in the discovery of the relation of one thing to another, by means of the relations they each bear to some third thing. This is effected through the juxtaposition of two propositions in such a manner that a third proposition is seen to follow necessarily from them. The essence of the process (and perception) exists in the word *therefore*, which expresses the "inference" and that its truth is a necessary consequence of what has been before laid down—the third proposition being certainly and necessarily true, if the two preceding propositions are themselves true.

These three propositions must not contain more than three terms. Each of the first two of the three pro-

---

* See *ante*, p. 260, and *post*, p. 309.

positions is called a *premiss*, while the third, which results from them, is the *conclusion*.

Such an arrangement of propositions is termed a *syllogism*, and its three terms have distinct names.

(1) The term which is the predicate of the conclusion is called the *major* term (A).

(2) That which is the subject of the conclusion is named the *minor* term (B).

(3) The third term is called the "*middle* term," because it is present in each premiss, but not in the conclusion.

The three propositions also have each a distinct name.

(1) One is named the *major premiss*, because it contains the major term.

(2) Another is the *minor premiss*, because it contains the minor term.

(3) The third is (as above said) called the *conclusion*, and it declares the relation of the major and minor terms to each other.

Thus in the syllogism:

        All men are mortal.
        Socrates is a man.
        Therefore Socrates is mortal.

"Mortal" is the major term because it is the predicate of the conclusion (*i.e.*, is asserted of the other term in the conclusion), and therefore the first proposition is the major premiss.

Similarly the second proposition is the minor premiss, because it contains the minor term "Socrates," which is the minor because it is the subject of the conclusion— *i.e.*, is the term of which the predicate "mortal" is asserted.

"Man" is the middle term, because it exists in both premisses and not in the conclusion.

The "major premiss" compares the major term with the middle term.

The minor premiss compares the minor term with the middle term.

The conclusion compares together the major and minor terms.

The validity of the whole process reposes upon the truth that, whatever can be affirmed or denied of a class of things (*i.e.*, of every member of it considered as a whole mass) can be affirmed or denied of each individual member which constitutes it.

The truth of this in each case is signified by the word "therefore"—expressed or understood.

There are ten Rules with respect to syllogisms.

(1) and (2) affirm that (as before said) there must be neither more nor less than three terms and three propositions.

(3) The middle term must be distributed at least once.

(4) No term must be distributed in the conclusion which was not so in the premisses.

(5) There must not be two negative premisses.

(6) If there is a negative conclusion there must be a negative premiss.

(7) If there is a negative premiss there must be a negative conclusion.

(8) There must not be two particular premisses.

(9) If there is a particular conclusion there need not be a particular premiss.

(10) If there is a particular premiss there must be a particular conclusion.

The four forms of categorical propositions (universal and particular as both affirmative and negative) we have* symbolised by the letters A, E, I, O, as denoting the quantity and quality of propositions.

Syllogisms are said to differ in *mood* when they differ with respect to the quantity and quality of their component propositions. Thus, AAA denotes a syllogism made up of three universal affirmative propositions, and IEO signifies one composed of a particular affirmative major premiss, a universal negative minor premiss, and a particular negative conclusion—"as Jones is a man, no man is without a backbone, therefore Jones is not without a backbone."

There are sixty-four possible ways of arranging three letters, and therefore there are sixty-four conceivable moods for syllogisms, but all these save twelve would, if used, sin against one or other of the ten rules just laid down.

Syllogisms are further said to differ as to figure—that is, according to the position of the middle term, and there are four such figures. These may be represented by letters which serve perfectly to represent valid syllogistic forms, whatever real terms may be substituted for such forms. If the premisses with such real terms be actually true, the conclusion necessarily will be actually true likewise.

The four figures may also be represented by the following diagrams, which portray the relative extent and relations of whatever may be represented by the terms.

---

* See *ante*, p. 291.

First Figure.—Here the middle term is the subject of the major premiss and the predicate of the minor:

> All M is A.
> But all B is M.
> Therefore all B is A.

FIG. 55.

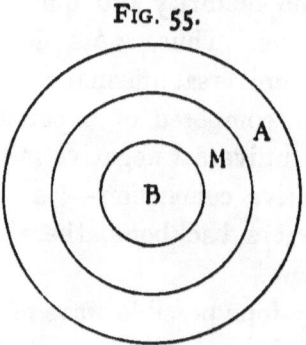

Second Figure.—Here the middle term is the predicate of both premisses:

> All A is M.
> But no B is M.
> Therefore no B is A.

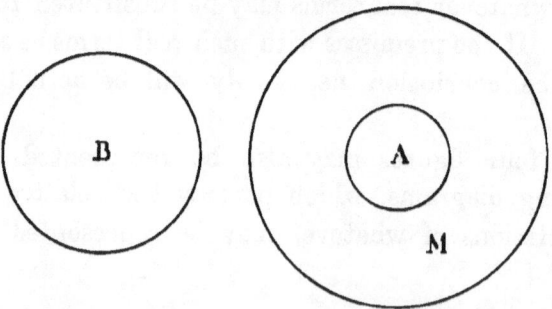

Third Figure.—Here the middle term is the subject of both premisses:

All M is A.
But all M is B.
Therefore some B is A.

FIG. 57.

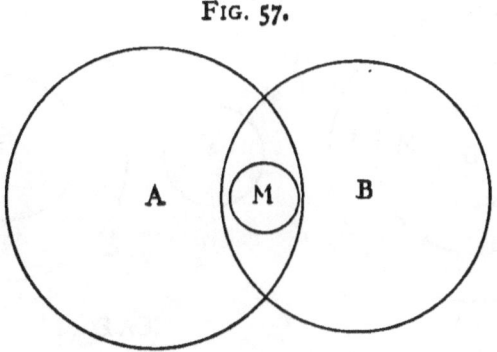

Fourth Figure.—Here the middle term is the predicate of the major premiss, and the subject of the minor:

All A is M.
But all M is B.
Therefore some B is A.

FIG. 58.

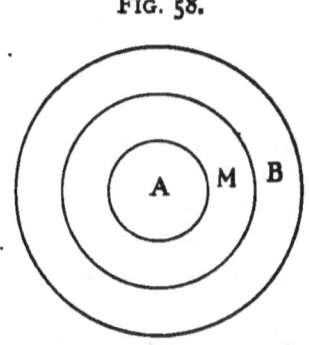

In the first figure the valid moods are: AAA, AII, EAE, EIO. AAI and EAO are also valid but superfluous, because the former is contained in AAA and the latter in EAE.

These four valid moods may be thus represented.

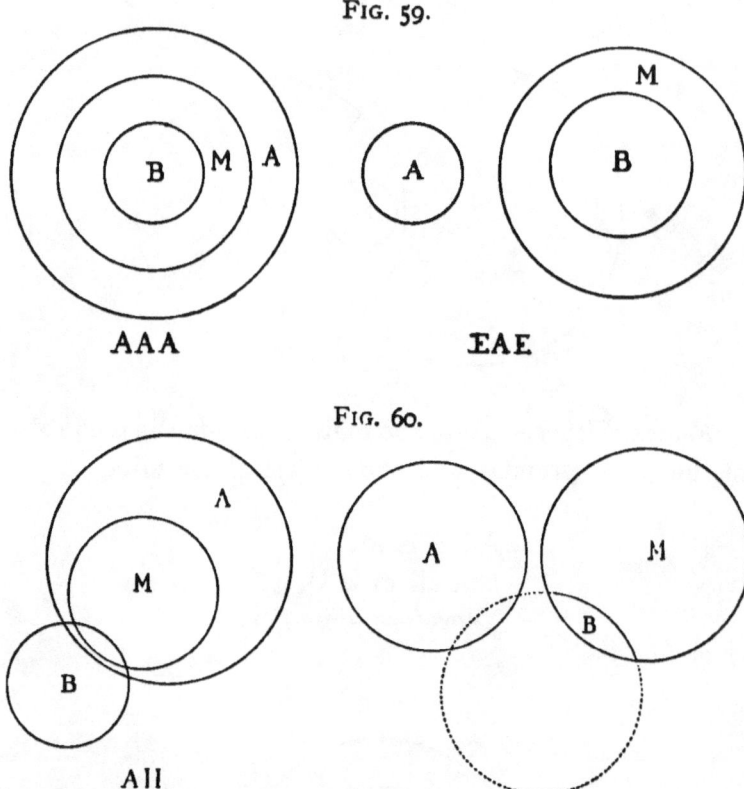

FIG. 59.

AAA   EAE

FIG. 60.

AII

The last of the four moods of the first figure will run:

No M is A.
Some B is M.
Therefore some B is not A.

There are two rules for syllogisms of the first figure.

(1) The major premiss must not be a particular proposition.

(2) The minor must not be negative.

Thus, if in the above syllogism the major premiss were "some M is A" the middle term would not be distributed in either premiss, since in the minor, M is only affirmed of some B and the totality of M is not regarded; therefore the third rule (p. 296) is violated.

Again, if the minor was negative the reasoning,

        No M   is   A
        (fish)   (feathered)
        Some B   is not M
        (animal)   (fish)
Therefore some B is not A
        (animal)   (feathered)

FIG. 61.

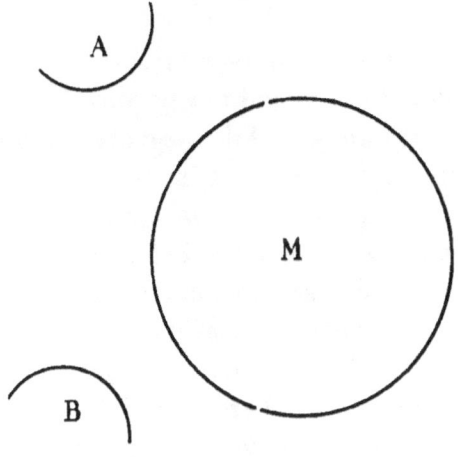

would be invalid, because A is distributed in the conclusion but not in the *major* premiss—a condition technically called "an illicit process of the major." It is evident that when we say "No M is A" we have not, and need not have, in view an extension of the whole periphery of A. What we say is that the whole of M has been surveyed and that none of it is A. But when we say "Some B is not A" we exclude some B from the *entire class* of A, and so A is distributed.

Various ingenious plans have been devised whereby syllogisms of the second, third, and fourth figures can be converted into syllogisms of the first figure. This process reposes upon those rules for the conversion of propositions which have been hereinbefore given (pp. 293 and 294). But for an account of such devices the reader is referred to special treatises on logic.

In the second figure, the valid moods are: AEE, AOO, EAE, and EIO. In this figure there must be a negative premiss, and therefore a negative conclusion. Thus:

> All apes are mammals.
> But no gill-bearing creatures are mammals.
> Therefore no gill-bearing creatures are apes.

If the second premiss were affirmative as well as the first, the middle term would not be distributed, as, *e.g.*, with such premisses as "All apes are mammals," and "Hair-bearing creatures are mammals," we can conclude nothing because the middle term "mammals" is distributed in neither case. For all that appears, there might be (as, in fact, there are) mammals which are neither "apes" nor "hair-bearing creatures."

In this figure also the major premiss must not be particular, as otherwise there would be a term distributed in the conclusion which was not distributed in the premiss, and this is against Rule 4 (p. 296).

The third figure has the following valid moods: AAI, AII, EAO, EIO, IAI, OAO. Here there must be no negative minor, but there must be a particular conclusion. Thus:

> All mammals have warm blood.
> But all mammals are air-breathers.
> Therefore some air-breathers have warm blood.

If the minor was negative, as, *e.g.*, "All mammals are not furnished with gills," we could not thence conclude that "Some creatures not furnished with gills have warm blood," because the term "furnished with gills" is distributed in the conclusion but not in the *minor* premiss, since there may be and are creatures "not furnished with gills" which are not "mammals." This would be an "illicit process of the minor."

Similarly, there must be a particular conclusion, as otherwise our conclusion would be "All air-breathers have warm blood," in which case the term "air-breathers" would be distributed in the conclusion, but not in the minor premiss, which only says that "All mammals are air-breathers," for all which it may be the case (as it is) that many air-breathers are not mammals.

The valid moods of the fourth figure are, AAI, AEE, EAO, EIO, IAI.

In this figure, when the major is affirmative, the minor must be universal; and when the minor is affirmative, the conclusion must be particular. Thus:

All apes are mammals.
All mammals have backbones.
Therefore some backboned mammals are apes.

If for the minor, we only had "some mammals have backbones," we could not know from our syllogism whether apes might not be mammals without backbones. In order that a conclusion in this figure should be universal, we require a negative minor. Thus:

All apes are mammals.
No mammals breathe by gills.
Therefore no creatures breathing by gills are apes.

The division of syllogisms into these figures is not a

mere useless trick of art; for one figure is more useful for some purposes than another, though generally the first figure is the best and most serviceable. It is more convenient, *e.g.*, to say "Our judges are not untrustworthy and corrupt," than it is to say "no men who are untrustworthy and corrupt are judges." If we wish to show how some men will labour earnestly to give just decisions without private reward, we may say:

 Our judges labour earnestly to give just decisions.
 Our judges will accept no private reward.
 Therefore some men who will accept no private reward will labour earnestly to give just decisions;

which is a syllogism of the third figure and the mood AAI.

But the utility of the rules concerning syllogisms may perhaps be made more evident by some examples of their violation. Thus:

 The temperate are healthy.
 Some ignorant people are healthy.
 Therefore some ignorant people are temperate.

This is a syllogism of the second figure and the mood AII, but there is no such valid mood, and in fact it is but a syllogism in appearance, for the middle term is not distributed in either premiss. It does not declare that all healthy people are either temperate or ignorant, and so, for all it says, there may be some (as in fact there are) who are neither the one nor the other. Again:

 Men of science are learned men.
 Men of science are long-lived men.
 Therefore all long-lived men are learned men.

This is a syllogism of the third figure and the mood AAA, but there is no valid mood of that kind in the third figure which needs a particular, not a universal, conclusion. The above apparent syllogism is invalid. There is in it an "illicit process of the minor," the term "long-lived men" not being distributed in the minor premiss, though it is in the conclusion.

A *Demonstration* is a syllogism, the major term of which denotes a "property," the minor the "species," and the middle term the definition* of a species. Thus:

    A rational animal is capable of laughter.
    Man is a rational animal.
    Therefore man is capable of laughter.

But all syllogisms are objected to by some people who affirm that they teach us nothing, because the conclusion of a syllogism only re-affirms what was already contained in its premisses. "Whoever has said," they repeat, that "all men are mortal" has already said in effect that "Socrates is mortal."

To test the force of this objection, let us see by an example what our meaning is when we declare that any one object belongs to a certain class of objects. Persons ignorant of zoology, may fancy that a whale is a fish. Nevertheless the whale in truth belongs to the class of beasts. Now when we make that statement what do we mean? We mean that a whale, in spite of its shape and exclusively marine mode of life, is nevertheless more closely allied in its nature to such creatures as cattle, beasts of prey, &c., than it is to any fishes. But even if we are zoological experts, we do not, in saying "a whale

---

* See *ante*, p. 288.

U

is a beast," distinctly advert in our minds to all those various anatomical conditions which characterise the class of beasts, but only to the fact of the predominance in the whale's organisation of the marks which distinguish that class of animals. We can, however, if we choose, turn back our mind, and mentally, or verbally, refer to any one of such marks or characters, and recognise the fact that the whale, inasmuch as it belongs to the class of beasts, must have that particular character so referred to—one of those various marks which are common to the whole class. Then we may say to ourselves, "The whale, being a beast, must have warm blood." We in this manner bring forward into *explicit* recognition a character, the existence of which was *implicitly* contained in the statement that it was a beast. This character may not only have never been previously thought of by us, but we may not have recognised the fact at all.

In repeating, then, the syllogism:

> All beasts have warm blood.
> The whale is a beast.
> Therefore the whale has warm blood,

a new fact may become explicitly recognised which previously was but latent, and so the syllogism can impart knowledge—it makes *implicit* truth *explicit*.

Now the difference between explicit and implicit knowledge is so great that the latter may not deserve to be considered "knowledge" at all. No one will affirm that a student by merely learning the axioms and definitions of Euclid,[*] will, by having done so, have also become at once acquainted with all the geometrical

---

[*] See *ante*, p. 34.

truths the work contains, so that he will have no need to study its various propositions, all of which he will thus know without having once read them. Yet all the propositions about circles, triangles, &c., in his "Euclid" are implicitly contained in the definitions and axioms. Although, then, a student may know that mass of geometric truths implicitly (in knowing the definitions and axioms) he does not for all that, really and actually know them. In order that he may learn truly to know them, he must go through those various processes of "inference" by which the different truths implicitly contained in Euclid's definitions and axioms are brought to the student's knowledge *explicitly*.

After this digression we will continue and conclude our notice of the elementary rules of logic.

One form of argument is termed a *Sorites*. In it a number of propositions are so heaped up together that the predicate of the preceding proposition is always taken as the subject of the following one, till the last predicate is enunciated together with the first subject.

Thus we may have:

> Peter is a man.
> All men are animals.
> All animals are living things.
> All living things are corporeal.
> All corporeal things are substances.
> Therefore Peter is a substance.

This reasoning really consists of a series of syllogisms, all the minor terms of which, except the first, and all the conclusions, except the last, are left out. All the majors, however, the first minor and the last conclusion, have been retained, while the retained minor and the first major have been transposed, the minor coming first. Thus, written out fully, the above sorites would become:

> All men are animals.
> Peter is a man.
> **Therefore Peter is an animal.**
> All animals are living things.
> **Peter is an animal.**
> **Therefore Peter is a living thing.**
> All living things are corporeal.
> **Peter is a living thing.**
> **Therefore Peter is corporeal.**
> All corporeal things are substances.
> **Peter is a corporeal thing.**
> Therefore Peter is a substance.

[The propositions in black type are the ones left out in the sorites.]

Another form of reasoning, in which there are at most two hidden, or cryptical, syllogisms, is called a *Prosyllogism.* Thus:

All tyrants, **because they transgress the law,** are unjust, but Napoleon I., **who violated his engagements to the Republic to make himself supreme,** was a tyrant, therefore Napoleon I. was unjust.

[The parts in black type are relics of two hidden or cryptical syllogisms.]

Thus written out at length the prosyllogism is:

> **All who transgress the law are unjust.**
> But tyrants transgress the law.
> Therefore tyrants are unjust.
> **All who violate their engagements to the Republic to make themselves supreme are tyrants.**

But Napoleon I. violated his engagements to
the Republic to make himself supreme.
Therefore Napoleon I. was a tyrant.
All tyrants are unjust.
Napoleon I. was a tyrant.
Therefore Napoleon I. was unjust.

Syllogisms may also be *hypothetical* instead of *categorical*, and hypothetical syllogisms may be *conditional* or *disjunctive*.

The following is their form:

(1) *Conditional:*

If A is B then C is D.
   (Antecedent) (Consequent)
or   if A is B then A is C
but   A is B therefore A is C.
or finally if A is B then A is C.
but if   A is D then A is B.
therefore if A is D then A is C

(2) *Disjunctive* syllogisms are thus expressed:
either   A is B or C is D.
or either A is B or A is C.

A *dilemma* is a particular form of " disjunctive " syllogism, *e.g.*,

If A is B then either C is D or E is F.
but neither C is D nor E is F.
therefore A is not B.

Induction * does not depend on the syllogistic principle, but on a belief in the order and continuity of nature.

---

* See *ante*, pp. 260 and 294.

It is true that the dictum "whatever you can assert of all things contained under a class, can be asserted of that class." This is manifest, obvious, and even trivial, and of no practical value.

The form of induction is as follows:—

$$a + b + c \text{ is } Z.$$
$$\text{but} \quad a + b + c = \text{all } X.$$
$$\text{therefore} \quad X \text{ is } Z.$$

It is, however, only in very rare instances that $a + b + c$ are *all* X.

The practical dictum of induction is: "What you can say of a sufficient number of particulars of any class under a required diversity of circumstances and conditions, you can fairly predicate of the whole class." It is the main process used in the development of the physical sciences.

It is the process of discovering laws from facts, and causes from effects.

$$a + b + c \text{ is } Z,$$

but $a + b + c$ though not all X may, from the peculiarity of their circumstances, be taken as practically = all X,

therefore X is Z.

Deduction, as we have now seen, derives facts from laws, and effects from causes.

We all notice, with more or less readiness and facility, resemblances and differences, and we group these in relation to their antecedents and consequences. As an example, we may take the fact before noticed [*] that while almost all bodies tend to fall, they fall differently in the open air and under the bell-glass of an air pump.

We study nature always by observations, but we can

---

[*] See *ante*, p. 60.

often so arrange the conditions of bodies as to make experiments with them. The latter process is especially useful in studying the cause of any action, or condition, of bodies.

The cause (or causes) of anything must be sought amongst its invariable antecedents or concomitants.

The circumstances in which all the instances in which anything takes place, alone agree, must be a cause. This is called the *method of agreement*.

If two instances have all their circumstances alike save one, and the event (the cause of which is sought) only occurs in that single case, then that one must, at the least, be closely related to its cause. This is the *method of difference*.

If two instances in which Y occurs have X, and two instances in which Y does not occur, have nothing in common but the absence of X, then X is the cause of Y. This is the *joint method of agreement and difference*.

If we subtract from a given effect, all that is due to certain causes, then the residue is the effect of the rest of the causes. This is the *method of residues*.

If X and Y increase, decrease, and vary together, then one is the cause of the other, or closely connected with such cause. This is the method of *concomitant variations*.

One effect may have several causes, therefore the method of agreement is uncertain, but not so "the joint method of agreement and difference."

A cause is generally suggested by analogy, or resemblance, from cases in which the connection of a cause and effect is better known.

A cause will, as a rule, be the more easily discovered, the greater the number of forms of the effect which are examined.

A suspected cause may be tested by allowing it to operate in circumstances of less complication, to see whether the effect will still be produced.

Such are the main practical rules which should be made use of when applying the inductive method in practice.

The chief part of logic, however, is the deductive system (the elements of which have been described in this chapter), which therefore, as before said, especially constitutes both the science of the laws of thought, and the art of reasoning.

CHAPTER IX

HISTORY

By History we mean an account of those successive conditions and inter-relations of different nations, which are based upon more or less trustworthy evidence. It includes an account of religion, speculative opinions, language, manners and customs, &c., as well as of migrations, wars, political organisations and the succession of different dynasties. This vast mass of information cannot, it is obvious, be sketched, even most briefly, in a single chapter—least of all in a chapter of such an elementary work as the present one. Various branches of inquiry must be here passed over entirely in silence, while others can be little more than hinted at, though the elements of historical science must not be altogether omitted. But the reader is referred to special treatises on the history of different nations, on philology, on manners and customs, on religion and on philosophy, for all that he may desire further to know on so extremely important a subject.

There are some persons who will not be content with a mere record, even a full one, of those conditions of mankind which are known to us through authentic records (written or engraved on stone), but demand an account of antecedent periods now commonly denoted by the term "prehistoric." Others go further still, and require a statement as to the probabilities respecting the origin of the whole human race. Such a question as this last, however, cannot be entered upon here, any more than the inquiry as to the origin of intellect * or of life,† or the absolute nature of light or heat.‡ The facts which have been herein brought before the reader's notice, are too elementary to warrant the introduction of statements which must be merely speculative. Persons specially interested in the question as to what light science throws upon the origin of man, are here referred to an antecedent publication of the present author.§ We may, however, even here, point out that since thought precedes language (whether of speech or gesture) it is impossible to understand how a rational being, one especially who has a perception of right from wrong, could ever have arisen from a basis of mere feeling, and from creatures incapable of abstract thought, even such thought as everywhere exists even amongst the very lowest savage tribes.|| However tempted we may be to assume it, it is also by no means evident that the most barbarous existing savages afford us the best attainable representation of the very earliest condition of mankind.

Of course when we take a wide view of all human history known to us, it becomes plain that, on the whole, there has taken place a process which we must recognise as one of progress and improvement; yet even with respect to savages, there is much evidence to show that very many of them have fallen from some higher antecedent condition. Social progress is an exceedingly complex phenomenon, the result of many factors. No

---

* See *ante*, p. 269. † See *ante*, p. 189.
‡ See *ante*, p. 86.
§ The "Origin of Human Reason." Kegan Paul, Trench, & Co., 1889. || See *ante*, pp. 264 and 268.

one will probably contest that in some matters of art, even we Europeans have fallen far behind our predecessors of more than two thousand years ago. The ruins which exist in Central America demonstrate the degradation which has there taken place, and the degradation of various tribes of American Indians is certainly known. The Fuegians and Caffirs also give evidence that they once possessed a higher social state than is theirs to-day. The world is, in fact, sown broadcast with the traces of higher social conditions and civilisations which have passed away, and bears many a scar due to the triumph of ignorance and brutality over relative refinement and culture.

Putting on one side, then, the questions which regards the very earliest condition of mankind, it has been ascertained that many races have passed through several prehistoric stages of existence. There was a time when the only weapons made use of by such races were roughly chipped, unpolished flint implements. These have been distinguished as *Palæolithic* men. Other and later races formed implements of ground flint, and they have been termed *Neolithic* men. Then we have races which made use of metal for attack and defence; first of weapons formed of bronze, on which account they have been called *Bronze* men, and later of iron, and such last have been named after that metal.

But as yet there is very insufficient evidence as to the length of time during which palæolithic men lived in any one region, nor has it been proved that palæolithic and neolithic men may not have co-existed in different regions, as we know "bronze" men and "iron" men did subsequently. But even palæolithic men could make drawings in outline of different animals, and we owe to them the only authentic representation of the mammoth, or extinct elephant, scratched on one of that animal's bones. For man had come into existence while various animals, now extinct, still inhabited the earth, amongst them, the mammoth, the woolly rhinoceros, the cave bear and the cave hyæna—so called because their remains have been found in caves,* as have the remains of many men.

Thus it is certain that the earth has had human inhabitants for a long period, although it is impossible as yet to express it with any accuracy in terms of years —save as a minimum. The civilisations of Egypt and of China extend back for more than six thousand years, but they were probably preceded by tens of thousands, possibly by hundreds of thousands, of years of unrecorded human existence. It may, however, be considered certain that man did not exist before the tertiary epoch, although it would be rash to deny the possibility of his having appeared in "Miocene" times.*

However that may be, many naturalists are now convinced that man's existence antedates the glacial epoch, and it is certain that he has witnessed great geographical changes and, as before said, the extinction of many animals. Some of these must have been formidable enemies, indeed, to men armed with no better weapons than were the ancient flint implements.

We must not, however, linger over matters prehistoric, but advance at once to note the elementary facts of history, and especially of that history which specially relates to the existing condition of our nation and of the nations which are most nearly related to us. Thus, for our present purpose, we may leave entirely on one side the

* See *ante*, p. 169.

inhabitants of all parts of the earth unknown to the government of the ancient Roman Empire, since all Europe is the outcome of Roman civilisation modified by the action upon it of the inhabitants of the provinces it conquered and of unconquered regions near it.

There appears much reason to believe that one stock, or early race, of mankind was the common parent of almost all the inhabitants of Europe, Persia and India, and this race has been distinguished as *Indo-Germanic* or *Aryan*. The exceptions are the Laps and Finns of Northern Europe, the Basques of the South-West, and the Turks.

The Non-aryan inhabitants of Asia were the Turks, Mongols and Chinese, together with the very remarkable people known as *Semites*—remarkable indeed because amongst them rose the Jewish, Christian and Mahommedan religions, while they also gave origin to some of the grossest and most cruel forms of paganism. The "Semitic" nations were the Jews, the Arabs, and the Phœnicians. The last named dwelt on the sea coast of Syria, whence they sent forth colonies, the most notable of which was Carthage, though some were beyond the Straits of Gibraltar, one of these being Cadiz. The inhabitants of Africa were of negro race, save the Egyptians in the north (where the valley of the Nile sheltered and, through its annually rising waters, nourished them) together with north-western tribes over which Carthage came to dominate.

Of the Aryan Europeans, the earliest to have authentic historical records, and the most famous in early science, literature and art, were the Greeks. It is one of the most wonderful facts of history that so small a people, inhabiting so restricted a territory of islands and peninsulas, mountains and valleys, should, almost unaided, have developed a civilisation which has ever since been the admiration of the most cultured of mankind, and which, in some respects, no moderns can equal, far less surpass. It was also developed with amazing rapidity, for from the giving of laws to Athens by Solon, B.C. 594, to its conquest by Philip of Macedon, was but a period of little more than two hundred and fifty years. The Greeks were divided into a number of small States, the most celebrated of which was Attica (whereof Athens was the capital), and the Peloponnesus, which consisted of what is now the Morea, with Sparta for its capital.

They early sent out colonies to Cyprus, Sicily, Southern Italy and the South of France (having founded there the city of Marseilles) and in many parts they came in hostile contact with colonies of Phœnicians, who were a people yet earlier civilised, hardy and indefatigable traders, from whom the Greeks seem to have learned their alphabet and who had colonised before the latter.

Those celebrated poems, the "Iliad" and the "Odyssey," though they must be deemed unhistorical, are yet thought to give us a tolerably faithful picture of early Greek life.

The governments of the different Greek States, each of which was the dominion of a city, at first consisted of an hereditary king, a council of chiefs, and a more general assembly of those who held the rights of citizenship. In Greece itself, the kingly power gradually disappeared to give way to an aristocratic government of privileged families (descendants of the oldest inhabitants) or to democracies—that is, government by the citizens assembled. These citizens, however, were by no means the same as the inhabitants of the city, but only a wider aristocracy. For in most cities there were many slaves and free men who were not natives. These had no power themselves, nor even their children, till they were made citizens.

Sparta retained the name of king for two of its hereditary magistrates, who exercised their power simultaneously. The two States north of what was at first alone deemed true Greece—namely, Macedonia and Epirus—retained the old kingly system of government.

Besides these forms of government, local disputes and disorders sometimes enabled an influential citizen to seize supreme power. Such a man was called a *tyrant*, and his system a *tyranny;* though such expressions did not originally denote the way in which such power was used, but only its form. The last notable tyrant in Greece itself was *Pisistratus*, whose sons were expelled about the end of the sixth century before Christ. In the colonies, however, especially in Sicily, they existed much later.

The expulsion from Athens of Hippias, son of Pisistratus, was the occasion of the first of the famous Persian wars. The Persians who, under Cyrus, in the sixth century B.C., took the great and powerful city of Babylon, also conquered a domain which included Greek colonies on the coast of Asia Minor, and thus became involved in disputes with Greece itself. Later, the Persian king, Darius, desiring to force the Athenians to take back Hippias, sent an expedition which landed in Attica, but was utterly defeated B.C. 490, in the celebrated battle of Marathon, by a relatively insignificant force of skilled and well-disciplined troops. Ten years later, the Persian king, Xerxes, came by land with a vast army, only to be routed at Thermopylæ, and soon the Persians were forced to withdraw for a time even from the Greek cities of Asia.

These efforts led to a great ascendency of Athens in Greece. It was a very democratic State, under the leadership (from 444-429 B.C.) of Pericles, who adorned the city with new temples and other public buildings, while the poets, Æschylus and Sophocles, produced their world-renowned plays, to be succeeded later by those of Euripides and Aristophanes. This also was the period of Socrates and Plato, of whom we have to speak later. The power and splendour of Athens increased till the jealousy of other great cities gave rise to the celebrated Peloponnesian war between Athens and aristocratic Sparta, with their respective allies. It lasted, with a brief interval, for twenty-nine years, and ended in the defeat of the Athenians at the naval battle of Egospotamos. But the subsequent predominance of Sparta was followed, after various struggles, by the practical subjugation of the whole of Greece under Philip, King of Macedon, whose son, Alexander the Great, as the leader of Greece, invaded and conquered Persia, received the submission of Egypt—where he founded Alexandria — and after penetrating into Northern India, died at Babylon, B.C. 323. He was the greatest conqueror the world had yet known, and the vast empire he erected became, after his death, divided between his generals. Thus his conquests gave rise to the dynasty of the *Ptolemies* in Egypt, and to that of the *Seleucidæ* in Syria (founded by two of his generals), and so the Greek language and Greek culture became spread over all the most civilised and important part of the then known world. At that time, Rome was only contending for dominion over adjacent Italian territories, and the Greek cities of South Italy still flourished in security and independence. Indeed, Greek influence and culture become more than ever diffused around the Mediterranean and notably in Western Asia,

Fig. 62.—The Dominions and Dependencies of Alexander. (*After Freeman.*)

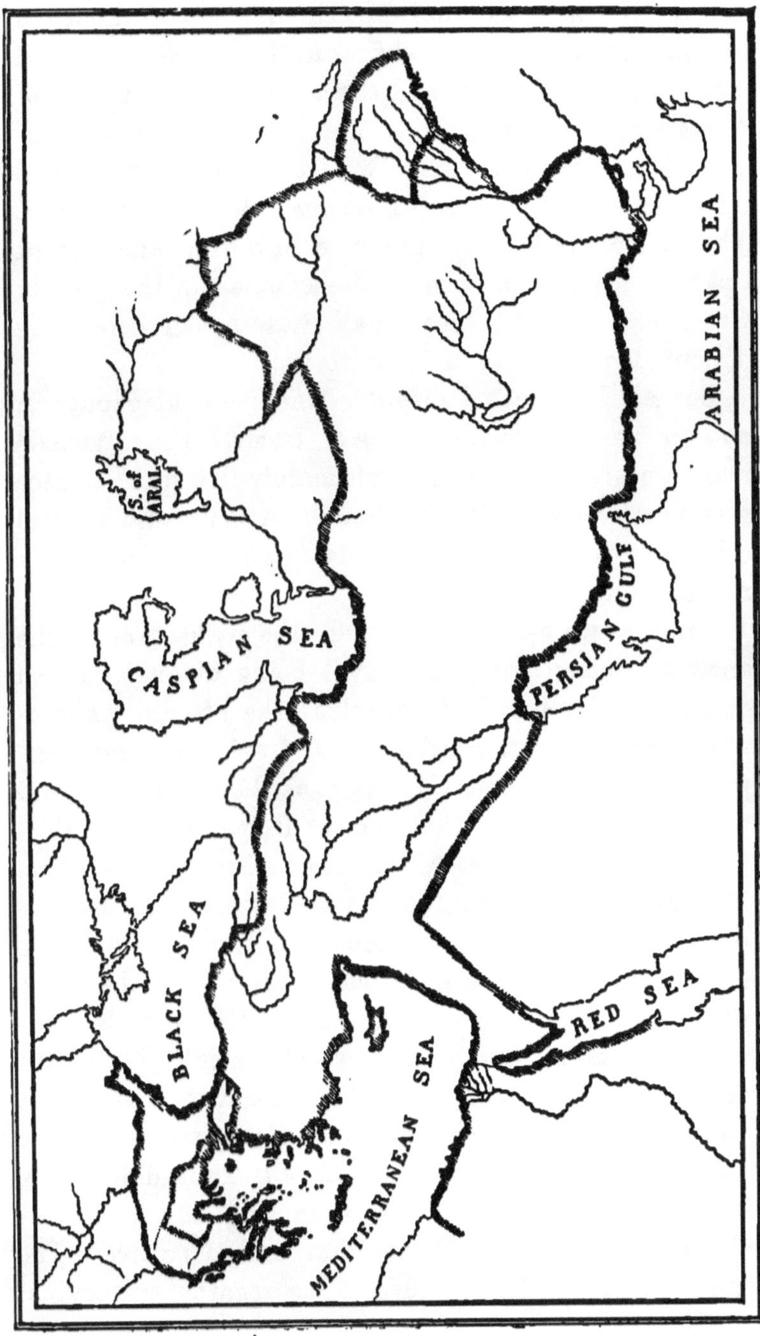

But Greece itself and Macedonia (now reckoned a part of it) fell into great confusion, the Macedonian royal family became extinct, and the crown passed to one of Alexander's generals, while another, Antipatros, conquered Athens B.C. 322. Epirus, however, rose into importance under King Pyrrhus, who died B.C. 272. The days of single city rule were now over, and, except Sparta on the south and Macedonia on the north, Greece was divided into small States, each consisting of confederated cities and territories.

But all internecine disputes were ere long put an end to by the growing power of Rome. Four Macedonian wars took place and ultimately (B.C. 146) Greece was practically subdued, though Athens and several other Greek cities and islands retained for a time a nominal independence.

Since, as we have already said, the civilisation of the ancient Roman Empire is still being directly carried down and furthered by ourselves, the artistic, philosophic and religious conditions of Greece are most important matters for consideration, as they so directly and powerfully influenced that ancient Roman civilisation.

The perfection of the plastic art in Greece, with which the names of Phidias and Praxiteles are associated, is revealed to us by many precious sculptured relics, while the ruins of the great Athenian temple of the city's patron goddess, Athene, with many others of less renown, proclaim the skill and refinement of Grecian architecture. Homer and the dramatists have been already mentioned, while, as historians, we may name Xenophon, Herodotus, and, above all, Thucydides, who, in his history of the Peloponnesian war, set a model for all time of historical description and narrative. The same, perhaps, may be affirmed, as regards oratory, of

Demosthenes; but for an account of Greek art and literature the reader must have recourse to special treatises. The reader should also seek elsewhere for a description of Greek religion and philosophy, though a few words on each of these subjects must here be said.

The religion of a very large part indeed of the human race has had its origin in the propitiation and worship of deified ancestors, though such marvellous natural objects as the sun, moon, and stars, doubtless in many instances received a self-suggested worship. Animals with their wonderful instincts, especially when taken in connection with human ancestors named after them, have been widely venerated, notably in Egypt, where bulls and crocodiles, cats and other animals, received divine honours in different localities. Even plants have not been altogether neglected, and, indeed, amidst tribes dwelling in almost treeless plains, such an object as a tree, self-sustaining and spreading abroad on all sides vast arms which give rise to, shed and renew, a multitude of leaves, may well seem the expression of some mysterious divine energy within it.

But the Greeks, with all the other divisions of the great Aryan family of nations, reverenced various powers of Nature as distinct divinities, whose names in different languages are akin. Thus, that of the old English god *Tue* (whence Tuesday) resembles Zeus, the great "cloud-compelling" god of the sky. The special god of the Greeks was indeed the sun, revered and worshipped as Apollo, who had, at Delphi, a world-renowned temple. The moon, the sun, and the dawn, had each their divine representatives; but such objects did not suffice to satisfy the Greek religious feeling. The actions of men, their passions and desires, were also deified, as in Ares, the god of war, in Aphrodite, of love, and in the god of wine, Dionysios.

But the Greek religion, like that of so many other nations, was especially a religion of cities, and, indeed, was of the essence of the city. For a city may be said to have been a collection of men adoring the same god by rites which were deemed effective for securing the exclusive or predominant aid and protection of the power so worshipped. Hence, to neglect the due worship of such a supernatural patron, was to be unfaithful to the interests of the city, and even to be more or less of a traitor. Any recognition of such a thing as a "right of conscience" justifying a dissent from established practice was impossible. It could not, indeed, well have been even conceived of.

Worship was made up of prayers, processions and mystic movements, the sacrifice of animals, and offerings of food, and various more or less precious objects, with incense, music, and song. But religion consisted in such acts themselves, and not at all in morality, save in so far as it was "moral" to do what was serviceable to the State. To do what had to be done, not for the pleasure of doing it, but because it was a thing religion demanded, had necessarily to some extent a moral character, even though the action in itself might have been an immoral one.

But the morality of Greece was exceedingly different from our own, above all as regards the relations of the sexes. This was doubtless in part the effect, and in part the cause, of the great multitude of fantastic and often gross fables which were related and believed concerning the actions of the gods worshipped. The Greek religion was not a religion of definite dogma, but of recognised ritual practices, and it did not profess to give any revealed account of the origin and essential nature of the beautiful world which Grecian eyes beheld.

Such explanations of Nature, as well as the inculcation of moral doctrines, belonged not to religion but to philosophy. Philosophy, after having studied the world, the nature of man and his duties, also occupied itself with attempts to penetrate divine things, the meanings of the popular tales concerning the gods, and the nature and attributes of the Supreme Being.

As to the explanations of the world which philosophy suggested, *Thales*, a man of Phœnician descent, born about 640 B.C., taught that water is the original source of all things; *Anaximander*, of Miletus (611 B.C.), affirmed that the principle of all things is an undefined matter, at once the source of what he deemed the elementary contraries—namely the warm, the cold, the moist and the dry. The earth, he said, arose from fluid, and all animals were at first aquatic. But he appears to have held that man has a soul of the nature of air.

Pythagoras (529 B.C.) founded a very influential society, and instituted numerous ethical regulations. He taught* that "number" was the principle underlying all things—in the word "one" is the beginning of them all.

Anaximenes, of Miletus (about 528 B.C.), represented air as the first principle, and fire, wind, clouds, water and earth, as having been thence produced by condensation.

Diogenes, of Apollonia, a contemporary of Pythagoras, followed Anaximenes in holding air to be the origin of things, but believed such air to be vital and intelligent, like the human soul.

Xenophanes (569 B.C.) taught that "the one" of

---

* See *ante*, p. 5.

Pythagoras must be one infinite self-existing intelligence.

Heraclitus, of Ephesus, identified the Supreme Divine Spirit with an ethereal fire, and affirmed that all things are in a flux and a perpetual "becoming."

Parmenides (515 B.C.), on the contrary, following Xenophanes, inculcated the necessary existence of "the one." He said that it alone existed and was everything, and that plurality and change were but empty appearances—mere deceptions of the senses obscuring the unchanging unity perceptible to thought.

Zeno, of Elea (who taught Pericles and was born about 485 or 490 B.C.), defended the doctrine of Parmenides, and denied the reality and possibility of motion, on the ground of four arguments, which readers will find described in histories of philosophy.

Anaxagoras (born about 500 B.C.) reduced all origin and decay to a process of mingling and separation, but assumed, as ultimate elements, an unlimited number of primitive determinate substances. They first existed mixed together, and then the divine mind, out of such chaos, formed the world.

Empedocles (of the same period) affirmed the existence of but four elements: earth, water, air, and fire, with love as a uniting, and hate as a separating force.

Democritus (born 460 B.C.) advocated the celebrated doctrine of atoms, declaring such atoms to be the invisible, intangible, primary elements which compose all things according to their different modes of combination and relative position.

The foregoing enumeration must suffice as a catalogue of the names of the principal philosophers of the first period of Grecian philosophy. Its second period is less directly concerned about the material universe, and deals by preference with man as a thinking being, his duties and his thoughts—that is, Ethics and Logic.

Certain teachers, of whom *Protagoras*, of Abdera (about 490 B.C.), is a type, accepting and applying the doctrine of Heraclitus "that all things are in a perpetual flux," asserted man to be the measure of all things, and that just as each thing appears to each man, so is it for him. Thus all things are, he taught, but relative; everything is uncertain in reality, even the existence of the "Immortal Gods." What applies to our perception of things, also applies to our perception of their relations, and thus all morality becomes undermined, because nothing can be certainly affirmed to be good save as it may appear to be good in the eyes of those who so regard it. Nevertheless this very moral revolt was an indirect assertion of the rights of conscience; for since that was certain to each man which seemed to him so to be, he had a justification for refusing to comply with behests he deemed wrong, and for denying the ethical validity of commands on the part of the State. Such teachers (since they were by profession instructors in eloquence and polite learning) were termed *Sophists*. In the teaching of later members of that school the evil consequences which attended their teaching became more conspicuous; as, *e.g.*, in *Thrasymachus*, who identified "right" with the personal interest of a man in power. A reaction against such doctrines was inevitable, and it came to a head in that immortal teacher *Socrates*, who was born about 470 B.C. His supreme claim to distinction and reverence consists in his having taught that virtue does not consist in acts but in intentions. He distinguished between merely external virtue and its true essence, which he affirmed to be absolutely dependent on moral perceptions. As our readers know, he was fatally misunderstood and condemned to die for not recognising the gods which the State recognised. Yet he was a devout man, and while dying from the cup of hemlock given him, reminded a friend that they owed the offering of a cock to the god Esculapius. His martyrdom for the cause of philosophic morality gave a wide-spread and most enduring influence to his teaching. In the Socratic principle, that virtue depends on knowledge, much of the future course of philosophy was indicated—namely, an examination of ethics on the one hand and of reasoning on the other.

Euclid of Megara (not the geometrician), being imbued with the notion of the school of Parmenides, that "the one" is everything, taught that "the one" is really "the good." He and his followers devoted themselves to that side of the Socratic teaching which concerned reasoning.

Antisthenes (born 444 B.C.) and his disciples devoted themselves to its ethical and practical side, which he morbidly exaggerated. But he appears to have affirmed the unity of God, despising the popular beliefs. From the place, named Cynosarges, where he met his followers, they acquired the surname of *Cynics*, of whom Diogenes has become the accepted type. He proclaimed the need not only of renouncing the luxuries but even the decencies of life, and adopted an ostentatious asceticism which in some of his followers became brutal and indecent.

Aristippus of Cyrene, a disciple of Socrates, laid stress above all on the exercise of the will according to reason. The Cynics sought for independence through the renunciation of enjoyment; but Aristippus advocated self-control in enjoyment, by which true pleasure was to be attained. That such is the real end of life and the test of truth was taught by him and his followers, who constituted the *Cyrenaic* school.

In *Plato* (from 427 B.C.) the diverse tendencies of the Socratic teaching were combined into a system. The highest good, he taught, to be neither pleasure nor knowledge, but the greatest possible likeness to God as the absolutely good.

Of his method, his theory of "ideas," his psychology and theology, nothing can here be said, except that his conceptions were in many respects so lofty, refined, and admirable, as to have largely been made use of by most esteemed Christian teachers at different times in later ages.

The same was the case to a still further extent with respect to *Aristotle* (born 384 B.C.), the tutor of Alexander the Great. He was perhaps the most wonderful genius the world has ever seen—his mental activity was so acute, profound, and manifold. His treatises not only on philosophy but also on ethics, politics, physics, psychology, and various other subjects, are all in various respects admirable. To him we also owe the science of logic, while he was the father of biological science, and many of his descriptions of animal anatomy are amazingly accurate.

Pyrrho (from about 360 to 270 B.C.) was the founder of the philosphical sect known as the *Sceptics*. His scepticism was so great as to include scepticism of his own system, and he declined to enunciate any affirmation. He and his school said: "We assert nothing, not even that we assert nothing." Thus it was necessarily impossible for them logically to even attempt to sustain their own cause.

As we have seen, the Cyrenaics affirmed that the greatest good was happiness. This was translated by *Epicurus* (341-270 B.C.) into pleasure—the sole good and true end in life to his followers, the *Epicureans*, who taught that the gods were also divinely happy, and never deigned to occupy themselves about human actions and affairs.

The greatest contrast to the Epicureans were the *Stoics*, a philosophic sect, which arose out of the school of the Cynics, being a refined and elevated modification thereof. Their founder, *Zeno* of Cyprus, alarmed at the spread of scepticism, taught publicly (about 308 B.C.) that the aim of man's existence is neither to be wise nor to enjoy, but to be virtuous, to which end both knowledge and pleasure are to be employed as means, while being kept strictly subordinate. The aim set forth was an eminently practical one—to become a "perfect man." Zeno was succeeded in turn by *Cleanthes* and *Chrysippus* (282-209 B.C.).

Meantime a fresh development of scepticism—called the *New Academy*—took place. It was promoted by *Arcesilaus* (315-241 B.C.) and later by *Carneades* (214-129 B.C.). This school did not adopt the absurd, self-contradictory scepticism of Pyrrho. It did not profess to deny that we can affirm anything with certainty, but only that we can so affirm nothing except appearances and our own feelings. Thus it admitted a certain kind of knowledge, but denied we could know anything otherwise than as related to ourselves and to other things. It was a doctrine affirming "the relativity of knowledge," and logically resulted in more uncertainty and scepticism than its upholders themselves admitted. Such were the main features of Greek philosophy in Greece.

But Grecian Egypt, at Alexandria, developed yet a third school of philosophy, known as *Neo-platonism*. It was essentially religious, and was largely influenced by the speculations of certain Jewish thinkers who had imbibed

Hellenic culture. But what it may be needful to remark about it must be deferred till we have noted various facts about those who so influenced it, and therefore we must now leave the Greeks and direct our attention to the Hebrews.

The Jews were a relatively small Semitic* people whose influence on the world has been even more profound and important than that of the Greeks, because it has affected not science, literature and art, but morals and religion. Very little need be said about their history in a work addressed to English-speaking people who are so familiar with the Bible. Much would have to be said, however, if there was here any need to consider and weigh questions which modern Biblical criticism has made matters of controversy; but in dealing with the mere elements of history it will be enough simply to refer to them.

When the reputed descendants of Abraham began and completed the conquest of Canaan, sacrifices were offered in various parts of the land, and by men who were not priests, as we read in the Books of Judges, Samuel, and Kings. Nevertheless they were worshippers of one God, Jehovah—whether or not he was regarded as more than a tribal deity who watched over the Jewish nation, as other nations were watched over and protected by their various gods. The prophets, Amos, Hosea, Isaiah, and Micah most probably date from before 623 B.C. They, and indeed all the prophets, defended the exclusive worship of Jehovah, insisted on his moral character, and proclaimed morality to be not only absolutely necessary, but that it is the chief element of all true worship.

In the separate Northern portion of the nation (Israel)

---

* See *ante*, p. 317.

Amos prophesied under its second king, Jeroboam, who died 743 B.C. The end of the conflict of Israel with Syria was, as the reader knows, the Captivity (725 B.C.) and loss of the Ten Tribes. By this loss the Southern nation (Judah) had its sentiments of patriotism and racial distinctness accentuated. A further concentration and intensification of the Jewish nationality was effected by the reform of King Josiah (623 B.C.), who promulgated the Book of Deuteronomy, whereby all sacrifice was limited to the temple of Jerusalem and a Levitical priesthood. This was at the beginning of the period of the conflict of the Jews with Babylon, which ended in the captivity of Judah, under Nebuchadnezzar (597-566 B.C.), a little before which were the writings of Jeremiah and Zephaniah, whose whole spirit is in complete accord with Deuteronomy. But with the captivity and consequent loss of independent nationality, the religion of the Jews was yet further intensified. The "nation" became more or less changed into an "hereditary Church," while the stream of prophecy began to flow in the direction of ritual observance, and was not, as before, almost entirely ethical. Thus the full development of Judaism was carried much further during the exile by Ezekiel, who was himself a priest and for years an official at Jerusalem. He commanded, in the name of Jehovah, that a distinction should be made between priests and Levites, the former alone being permitted to sacrifice. Divine rewards and punishment for the individual came to the front in place of the older notion which regarded the community almost exclusively. Instead of what was said by the old prophets: "Let Israel love justice, and Jehovah will reward the nation accordingly," Ezekiel taught: "Let the individual love justice, and Jehovah will reward him."

After seventy years captivity, Cyrus, having conquered Babylon, permitted many of the Jews to return (536 B.C.), under Zerubbabel, and they began to rebuild the temple. Under Darius (520 B.C.)—the period of Haggai and Zechariah—and under Ezra and Nehemiah (456 B.C.), of whom Malachi was a late contemporary, the return from exile was completed. Then, it is now supposed, the Books of Chronicles were written and the Levitical code completely elaborated, and it is to this period that modern critics attribute the Pentateuch, which incorporated many very ancient fragments of divers origins and authorships.

Under the mild rule of the Persian kings, the Jews were allowed to manage their internal affairs, and the high priest was their chief magistrate. They strictly avoided inter-marriage with foreigners, and their detestation of idolatry became extreme, as also did their spirit of nationality and attachment to their peculiar observances. The nation thus lived peacefully, till, in the year 333 B.C., Alexander the Great appeared in Syria, when Jerusalem made its submission and was spared. Upon the death of the conqueror, Judæa fell under the dominion of the Ptolemies, who favoured the Jews, and planted a colony of them in Alexandria which had much influence on the intellectual and religious future of the world, as we shall see. But a general named Antiochus, who accompanied Alexander to the East, had a son Seleucus, who ultimately obtained the sovereignty of Syria, raised the city of Antioch and founded a dynasty—that of the Seleucidæ.*
Their kingdom extended at first over the eastern part of Asia Minor, and thence to beyond the Indus. About 256 B.C., however, a people of Northern Persia—the

---

* See *ante*, p. 320.

Parthians—revolted and established a powerful kingdom between the Caspian and the Persian Gulf. Under the Ptolemies and Seleucidæ, Greek influence gained much ground even amongst the Jews, while, as before said,* it then extended far and wide around the Mediterranean. Judæa passed under the sway of the Seleucidæ during the reign of Antiochus the Great—198 B.C. He visited Jerusalem, confirmed its privileges, and by his benignity still further promoted the influence of Hellenic culture over the Jews. But the increase of that influence greatly augmented the national and religious passions of such of the Jews as did not yield to it, so that a strong antagonism was produced between these two sections of that nation.

In the reign of his second son, Antiochus Epiphanes, a great change took place. He determined to force the Jews to renounce Jehovah and worship the Roman god Jupiter. Intrigues to obtain the post of high priest and consequent disorders, for a time favoured his attempt, and, after great slaughter, the Jewish worship was abolished and sacrifices to Jupiter were offered before his image in the temple of Jerusalem. This gave rise to the well-known rising of the Maccabees, which, after an alliance with Rome, succeeded. Ultimately the Maccabee Jonathan became king of the Jews, and founded a dynasty—that of the *Asmonœans*—which lasted for about a century till the last of the dynasty was killed by Herod the Great, who, with the support of the Romans, became king of Judæa 38 B.C., which at his death became a district of the Roman province of Syria. A succession of exceptionally rapacious Roman governors led to the terrible revolt which ended in the total

---

* See *ante*, p. 320.

destruction of Jerusalem, 70 A.D. One more struggle produced a further desolation of Judæa by the Emperor Hadrian, who issued an edict forbidding circumcision, the reading the Mosaic law (the Pentateuch), and the observance of the Sabbath.

Then the dispersion of the Jews over the world became greatly augmented. It had indeed begun much earlier, under the Ptolemies, and was, as before said, promoted by Antiochus the Great, who settled large numbers in various Asiatic cities. In the time of Cicero there was already a wealthy community of Jews in Italy. After Hadrian, the Jews spread throughout the Empire were allowed to follow their old usages and rites, and new synagogues and schools were erected in all directions. But they strictly maintained their racial distinctness and remained a people apart and by themselves, in a way quite different from any other people—Egyptians, Greeks, or Gauls—which had their several representatives in various parts of the Empire. This fact is very characteristic of the Hebrews, but most important of all is their strict monotheism and their detestation of idolatry, whereby they had come to differ so exceedingly from all other nations of that period of the world's history, and most notably from their brother Semites, the Phœnicians of Syria, and the Carthagenians. Part of their spirit of nationality was due to, as it was intensified by, the various prophecies which had led them to expect (though they rejected) the Messiah, and which had caused various impostors or enthusiasts to assume to themselves that character during the century which preceded the destruction of Jerusalem.

The Jews had done nothing to advance either science or art, and did not do so for centuries, nor till they came closely in contact with the Greeks at Alexandria, did they influence the course of philosophic thought. Their great characteristic was their religion and the lofty moral precepts which were closely interwoven therewith. In spite of a tendency slavishly to adhere to minute ceremonial observances, their moral atmosphere was a healthy one indeed compared with that of the Greeks. Their ethical precepts were not confined to speculative exceptional minds—the disciples of some philosophic school—but were disseminated far and wide throughout all classes of the people. In their firm determination rather to undergo persecution than violate their conscience as regards their religious duties, the Jews set a great and novel example to the world. Many were the martyrs who thus suffered under Antiochus Epiphanes and subsequently under the Roman emperors. Torture and death were not seldom their portion as a consequence of their refusal to perform what in their eyes were acts of detestable idolatry.

In speaking of the Jews it has been necessary to refer to the Romans. We will conclude this chapter with an elementary notice of Roman history, which will bring us to the dawn of the modern world, and enable us to see those factors from the interaction of which our present civilisation has arisen as a necessary consequence.

The *Aryans* who entered the peninsula of Italy, found it already inhabited, probably by a race akin to the existing Basques. The first Aryan swarm seem to have been the Celts, who, besides appropriating what is now France and the British Isles, occupied all the north of Italy down to where now stand Milan and Bologna. This latter was known as *Cisalpine Gaul*, the Celtic continental dominion north of the Alps constituting *Transalpine Gaul*.

The *Teutons* (the ancestors of the English, the Germans,

and the Scandinavians) and the *Slavs* (the ancestors of the earliest Prussians, the Poles, Russians, and Bohemians) were the two great swarms of Aryans who established themselves in Europe.

Italy at and near the coast, from the Po to what is now Genoa, and thence for a space westward, was the province of the *Ligurians*, while a more or less closely allied race peopled the islands of Corsica and Sardinia, with part of Sicily. The migrations, or invasions, which introduced what, in a restricted sense, are to be called *Italians*, have not yet been ascertained. Another people (settled in Italy before the dawn of history) were the *Etruscans*—renowned for their cities, constructed of large stones, and their superstitions. History discovers them with a thriving domain still extending over the Eastern half of Italy from the Arno to the Tiber. Around quite the south of Italy were a number of Greek colonies, some of which so much extended and prospered as to cause that part of Italy to be distinguished from what was the mother country of its inhabitants, as *Great Greece*. The rest of Italy—between Etruria, Cisalpine Gaul and Magna Græcia—was inhabited by the Italians, who seem to have been a race more nearly allied to the Grecian than to any other branch of the great Aryan stock. Nevertheless they were not, in early times (like the Greeks), a seafaring and colonising people, a fact probably due to the very much less indented coast-line of the country they inhabited. The Italians consisted of various tribes, whereof, from the west coast inwards, the north was inhabited by the Umbrians; while further south were the Apulians, east of which were the Samnites, who reached down towards the Lucanians—the neighbours of the Italian Greeks. Between the Umbrians and what ultimately became

Y

Rome, were the Sabines, south and south-east of these were the Latins and the Volscians, while between the Latins and the Umbrians were the Æquians.

The origin of Rome is still unknown, as is the history of its early kings. Situated on the Tiber, which was the south Etrurian boundary, it was probably erected as a bulwark of the Latins to defend them from the Etrurians. It was built on a hill known as the *Palatine*, where was a settlement of the Latin tribe called *Ramnes* or *Romans*. Another settlement, on the Capitoline hill, was made by the Sabines, and a league was established between the two; so that they formed one city with two sets of inhabitants—the Sabines becoming reckoned as *citizens* of the first settlement and therefore "Romans" also. Their kings were not hereditary, and the last of them, the Tarquins, have been thought to have been of Etruscan origin. They much adorned Rome, and made it the most powerful of the Latin cities. The monarchical constitution is not precisely known, but the people consisted of two classes—the *patricians*, or nobles, and the *plebeians*, or common people. The former were probably descendants of the oldest inhabitants, and the latter of men admitted later to the city's privileges. From these classes there arose (1) a noble assembly or *senate*, and (2) a popular assembly. The plebeians were full Roman citizens, but there was a gradually increasing mass of slaves who had no rights whatever; so that the plebeians formed a sort of secondary aristocracy of the whole population of the city. About 510 B.C. the monarchy was put an end to through the king Lucius Tarquinius—called *Superbus*, or proud—being, with his family, expelled from Rome. A highly aristocratic republic was then established, with the senate and popular assembly as before, but also with two annually elected magistrates, called *consuls*, at the head of the State.

Thereafter great dissension arose between the plebeians and the patricians, which latter had kept to themselves all the important offices of the State and were very oppressive. About 460 B.C. the people obtained the appointment of two officers, termed *tribunes*, whose office was to protect the plebeians against the patricians. But space cannot here be afforded for describing the Roman State and its officers, for a knowledge of which and of all the details of later disputes and dissensions, the student is referred to works on Roman history. We must here observe, however, that gradually a reconciliation was effected, and a consul, *Lucius Sextus*, was elected from the plebeians, 366 B.C. Before then, Rome had to struggle with her near neighbours the Etruscans, the Æquians, and the Volscians, fortifying herself by effecting a close alliance with the other Latin cities united together in a league. In time of danger, a special officer, called a *dictator*, was chosen, to whom great power was accorded for six months. Such a dictator was *Marcus Furius Camillus*, who (396 B.C.) captured the nearest Etrurian city (Veii) and greatly extended the Roman power. Six years later, however, they suffered from an invasion of the Cisalpine Gauls, who even captured the city, though it was soon again liberated. This was but one of several Gaulish inroads, while wars went on with surrounding tribes which ended in a repeated small increase of Roman territory—their conquered inhabitants being admitted to the privileges of Roman citizenship.

More serious wars, for the subjugation of all Italy to the city of Rome, then followed—wars with the Latins themselves, the Samnites, and others—till (by 282 B.C.) Rome had subdued almost all Italy, save some of the cities of what had been *Magna Græcia*. The populations thus subdued were divided into three categories. (1) Those to whom the great privilege of Roman citizenship was conceded; (2) those to whom a few municipal privileges in connection with the Government of Rome were granted. Such were first bestowed upon the Latin cities, on which account it has been termed the *Latin franchise;* (3) those known as "Italians," who were merely *allies*, having no municipal privileges (save their own independent ones), but being bound to follow the lead of Rome in matters military.

The next objects of attack were the Grecian cities of the south, beginning with Tarentum, which applied for help to Pyrrhus, king of Epirus, who came and was joined not only by the Italian Greeks, but also by some of the more lately conquered, true Italian people. After some success he was, however, defeated at Beneventum (276 B.C.) and had to leave Italy. A few years later, Rome subdued the whole of Southern Italy.

The time had now come for the powerful and warlike Roman republic to enter into contest with States altogether external to Italy, and first with the neighbouring State in Africa.

Carthage was a very wealthy, powerful, and influential city long before Rome began to attempt the subjugation of Italy, and a treaty was made between the two cities when the regal government of Rome came to an end. Carthage was governed by a senate which does not seem to have formed a close aristocracy, and the extent of its commercial expeditions and conquests no doubt largely contributed to greatly develop the intelligence of the community which peopled what seems to have been a generally tranquil and well-governed city. As early as 490 B.C. Darius is said to have in vain solicited its assistance in his invasion of Greece. It had acquired extensive dominions across the sea, having obtained possession of both Corsica and Sardinia, and a large part of Sicily, long before Rome had acquired any transmarine territory. In the last-named island, Carthage had to maintain frequent contests with the Greek inhabitants, whom Pyrrhus aided during a certain interval in his contest with Rome, before mentioned.

Roman soldiers were Roman citizens, but Carthage relied mainly on the aid of mercenary troops hired in Spain, and even in Gaul, as well as in Africa.

Sicilian disputes involved Rome in a contest with Carthage for the sake of protecting certain Italians who had taken possession of Mycenæ. Thus arose the *first Punic war*, which lasted from 264-241 B.C., when Carthage sued for peace and yielded up her Sicilian territories to Rome. Thus it was that Rome acquired her first foreign dominion—her first *province*—which consisted of all Sicily save what belonged to Heiron, the Greek king of Syracuse.

Thereafter Rome added to her dominions the Carthagenian islands, Corsica and Sardinia, while Carthage was acquiring large dominions in Spain through the military genius of Hannibal, son of Hamilcar. At last he took the city of Saguntum, which Rome claimed as her ally, and so began the *second Punic war*, which lasted from 218-202 B.C. During it, Hannibal made his celebrated march across the Alps, defeated the Romans at the battle of Cannæ, and raised several of Rome's allies in revolt against her. But the Romans meanwhile conquered not only the Syracuse kingdom in Sicily, but also the Spanish dominions of Carthage; while, finally, a naval battle was gained, under the Roman commander Scipio, the result of which was that Carthage had to

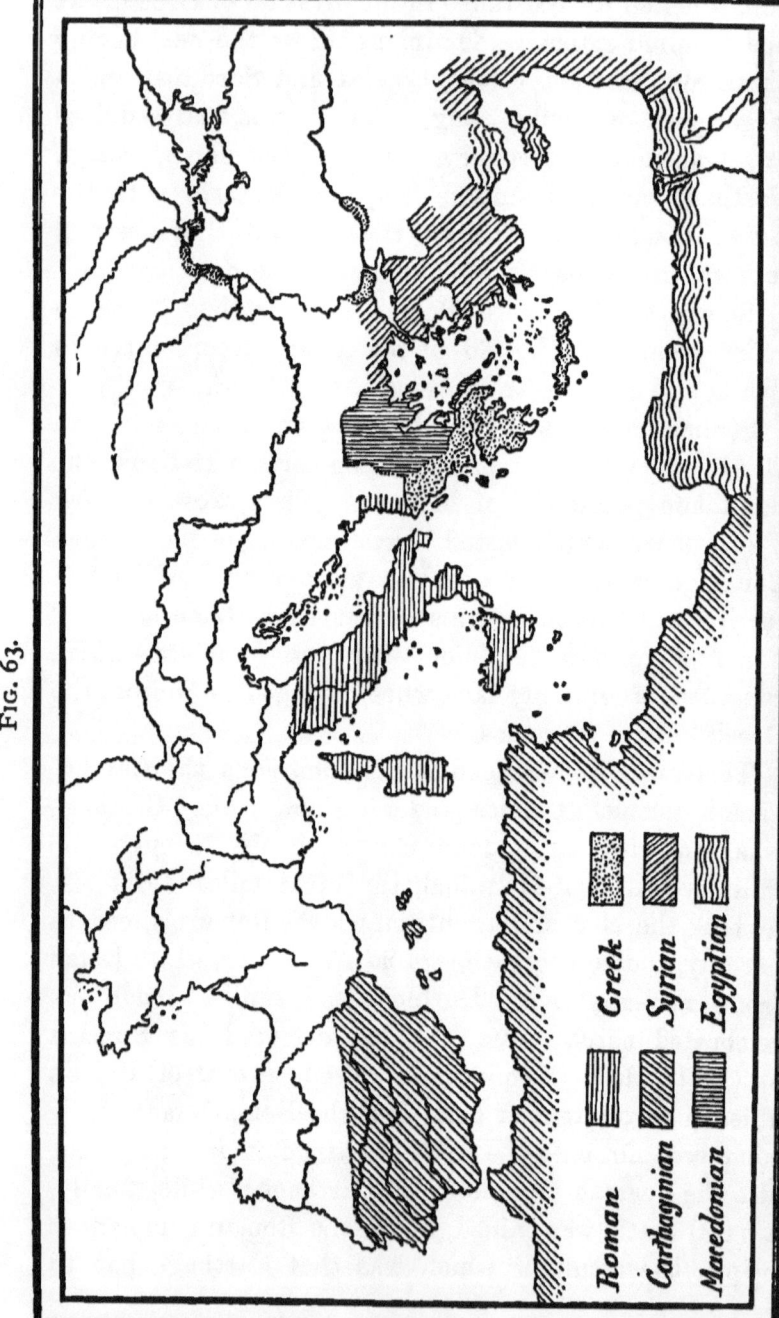

Fig. 63.—THE MEDITERRANEAN LANDS AT THE BEGINNING OF THE SECOND PUNIC WAR. (*After Freeman.*)

give up all her African conquests, and become a dependent ally, bound not to make war without Rome's consent.

Half a century later, the sovereign of the African kingdom of Numidia (which had been freed from the dominion of Carthage), who had been an ally of Rome in war, became at variance with Carthage, and this led to the *third Punic war*, which, after a struggle of only three years, ended in the destruction of the great African city (146 B.C.) and the reduction of its territory to the condition of a Roman province.

Meantime, during the second Punic war, the Romans had begun a strife which ended in the subjugation of Macedonia and Greece, which, with the exception of Athens and a few other cities, was completed at the same epoch as that wherein Carthage was destroyed. During these Macedonian wars, Antiochus the Great* crossed the Ægean Sea (192 B.C.) to aid the Greeks, but was driven back by the Romans, who followed and again defeated him. The result was that his dominions, which had previously extended from the Ægean to far beyond the Tigris, were reduced to Syria, with Antioch for its capital, the conquered country being divided between Rome's allies, whereby the Romans became the real masters of Western Asia.

The Cisalpine Gauls had greatly aided Hannibal in his invasion of Italy, in consequence of which they were likewise subjugated (about 191 B.C.), as also Liguria and the Venetian territory, so that all we now call Italy became then subject to Rome.

In the third century B.C. Spain was almost entirely peopled by Basques (called Iberians), save that Celts

---

* See *ante*, p. 334.

had penetrated to its centre and that there were Greek and Phœnician colonies on its coasts. By 218 B.C., however, the southern half of Spain, with the Balearic Isles, had been conquered by Carthage. But these conquests, by 206 B.C., had passed to Rome, and by 130 B.C. all Spain was a Roman province, save its northern mountainous region and Gallicia.

In 125 B.C. a Roman province, now *Provence*, was formed in Transalpine Gaul and a colony founded at Aix, while twenty years later the Roman domain reached northwards to Geneva and Toulouse. Any further advance was for a time stopped by a prodigious invasion of northern tribes—known as the *Cimbri* and *Teutones*—who were ultimately routed by the Consul *Marius*.

While this enormous extension of the power of the Roman citizens took place, there was at first no parallel extension of their number, for the *provincials* and *allies* had no power or voice in the Government, although Rome itself was becoming more democratic by the abolition of patrician privileges.

The sovereign power was in the hands of an assembly of the people which passed laws and elected magistrates, thus practically electing the Senate, which mainly consisted of men who had filled office. But there was a much worse distinction than an aristocratic one—that between rich and poor—with which co-existed, manhood citizen suffrage, and thus the sovereign assembly of the people became a turbulent or venal mob. The old Roman citizens grew scarcer from the slaughter of war, while strangers who had been slaves but were emancipated, came, as *freedmen*, to take their place. The dissensions which ensued are described in every history of Rome, and amongst them was the so-called sedition which ended in the slaughter of the two brothers named *Gracchus*—in 133 and 123

B.C. Thereafter Marius (the before-mentioned conqueror of the Cimbri) espoused the cause of the poorer classes and of the Italians who desired to be admitted to Roman citizenship. Many had taken up arms (90 B.C.) and, on their submission, obtained their desire. The Samnites, however, would not submit, and then *Sylla*—who was sent with Marius against them—attained great credit at Rome, so that violent jealousy arose between them, which led to the first civil war in Roman history. Sylla was called away to chastise Mithridates, king of Pontus (the region south of the Euxine), who had massacred Romans and invaded Greece. Sylla stormed Athens, defeated Mithridates, and returned hastily to Rome, where he assumed supreme power as *perpetual dictator*, and exterminated the party of Marius. At the end of the war all the Italians obtained Roman citizenship, save the almost annihilated Samnites. In 74 B.C. Mithridates made war again, and having been defeated by the great General Pompey, his kingdom was subdivided. Pompey then chastised the Pontic king's allies, erected in Syria a Roman province, and took Jerusalem, 63 B.C., Palestine then remaining under tributary kings as before mentioned.

The disorder at Rome, and the conflicts between its leading men, had become extreme, and while many desired to maintain the aristocratic republic which had so long existed, certain clear-sighted men saw that it was no longer in a condition fit to rule over such enormous and varied dominions.

Amongst those who strove to maintain their existing political state were *Pompey*, a greatly esteemed citizen *Cato*, and the world-renowed orator *Cicero*.

Another and most gifted Roman had married the daughter of Marius, and though of an old patrician house, espoused the popular side. He did this the more readily because his many gifts enabled him to control the Roman mob. This man was *Julius Cæsar*, who visited Britain and conquered the whole of Gaul from the sea to the Rhine.

During this time the Roman Government became more disordered than ever. Then Cæsar rebelled (49 B.C.) and marching on Rome got himself chosen Consul and Dictator by the people. He accumulated offices, becoming also *Pontifex Maximus* or High Priest of Rome, and assumed the title of *Imperator* (Emperor or Commander)—a military appellation which he made exclusively his own. After defeating Pompey in Epirus, and his other enemies in Spain and Africa, he was, as the reader knows, killed in the Senate House by Cassius, Brutus, and others, March 15th, 44 B.C.

Thereupon followed another civil war between Cæsar's partisans, under his great-nephew and adopted son Octavius, and Mark Antony—one of Cæsar's officers—on one side, and the supporters of the republic on the other. The battle of Philippi (42 B.C.) in Macedonia crushed the latter.

Then the triumphant Mark Antony fell, through love, to be the creature of the Egyptian Queen Cleopatra, and these, being defeated by Octavius, put themselves to death, when Egypt became another Roman province. Octavius, who had been adopted by Julius, was also Cæsar, and a new title, Augustus, was given him by the people of Rome (27 B.C.). Thus he is known as the Emperor *Augustus*, and he was reigning at Rome when Christ was born in Judæa. Although Augustus became an absolute sovereign, he assumed no royal pomp, appearing but as a citizen who was the first and highest magistrate, while the old republican institutions and forms continued to exist. He completed the subjugation of Spain, and of all that was previously unconquered south of the Danube. His stepsons, Tiberius and Drusus, with the son of the latter (Germanicus), also endeavoured to subdue the Teutons east of the Rhine, penetrating to the Elbe. The Romans were, however, driven back by the German Arminius (9 A.D.) and never obtained any permanent hold on the land east of the Rhine and north of the Mayn.

The great distinction of the Romans, apart from their military prowess and their political genius, was their instinct for legislation. They may be called the originators of jurisprudence; and their legal system, which began to be elaborated half a century after Augustus, has had an enormous influence over mediæval and modern Europe, and its influence still exists. Cicero has been already mentioned, and many of our readers doubtless are acquainted with his works. Julius Cæsar was also hardly less celebrated as a writer than as a soldier. The memorable poet Lucretius was a contemporary of Cicero. The familiar expression, "an Augustan Age," however, originated from the fact that the time of the second Roman Emperor was that of the most celebrated of Roman poets—Horace, Virgil, and Ovid, as also of the historian Livy.

Augustus reigned forty-one years, and was succeeded by his adopted stepson, the cruel and jealous *Tiberius* (14 A.D.), who put to death all those he feared, and was succeeded (37 A.D.) by *Caligula* (grandson of Drusus) who, after four years of mad wickedness of all kinds, was killed by his soldiers, who elected his uncle *Claudius* in his place. He was the first emperor chosen by the army, but its choice obtained the ratification of the Senate. Now began the conquest of Britain, which was visited by the Emperor (43 A.D.) He was poisoned by his wife and niece, Agrippina, and was succeeded (54 A.D.) by *Nero*, whom Claudius had been forced by his wife to adopt. After beginning well, he became one of the most horribly corrupt of all emperors, even causing his own mother to be slaughtered. Having been deposed by the Senate after a reign of fourteen years, he died by his own hand 68 A.D.

Thereupon ensued a period of conflict and confusion, emperors being chosen by the military in different regions, some of them for a time obtaining possession of Rome and legal recognition, as did *Galba*, *Otho*, and *Vitellius*. It was during and after this period that the poet and satirist Lucian lived.

A more settled state of things was initiated (70 A.D.) by the choice of *Vespasian*, by whom, in conjunction with his son *Titus*, Jerusalem was destroyed. The latter succeeded his father, 79 A.D. but died after a reign of but two years, universally regretted, to be succeeded by his brother *Domitian*, a most horrible tyrant, but who was not killed till 96 A.D. In his reign the Roman general Agricola completed the conquest of England, and war began with the *Dacians*—a people who inhabited what is now Hungary east of the Theiss, with Roumania and Bessarabia, and an adjacent slip of what is now Russia. It was in the year of the accession of Titus that the celebrated man of science, Pliny, met his death by incautiously approaching too near to Vesuvius at the time of that eruption which destroyed Pompeii, Another most celebrated historian of this period, to whom we owe a description of the Germans, was Tacitus.

After the two years reign of the Emperor *Nerva*, *Trajan* became emperor (98 A.D.), and reigned till 117.

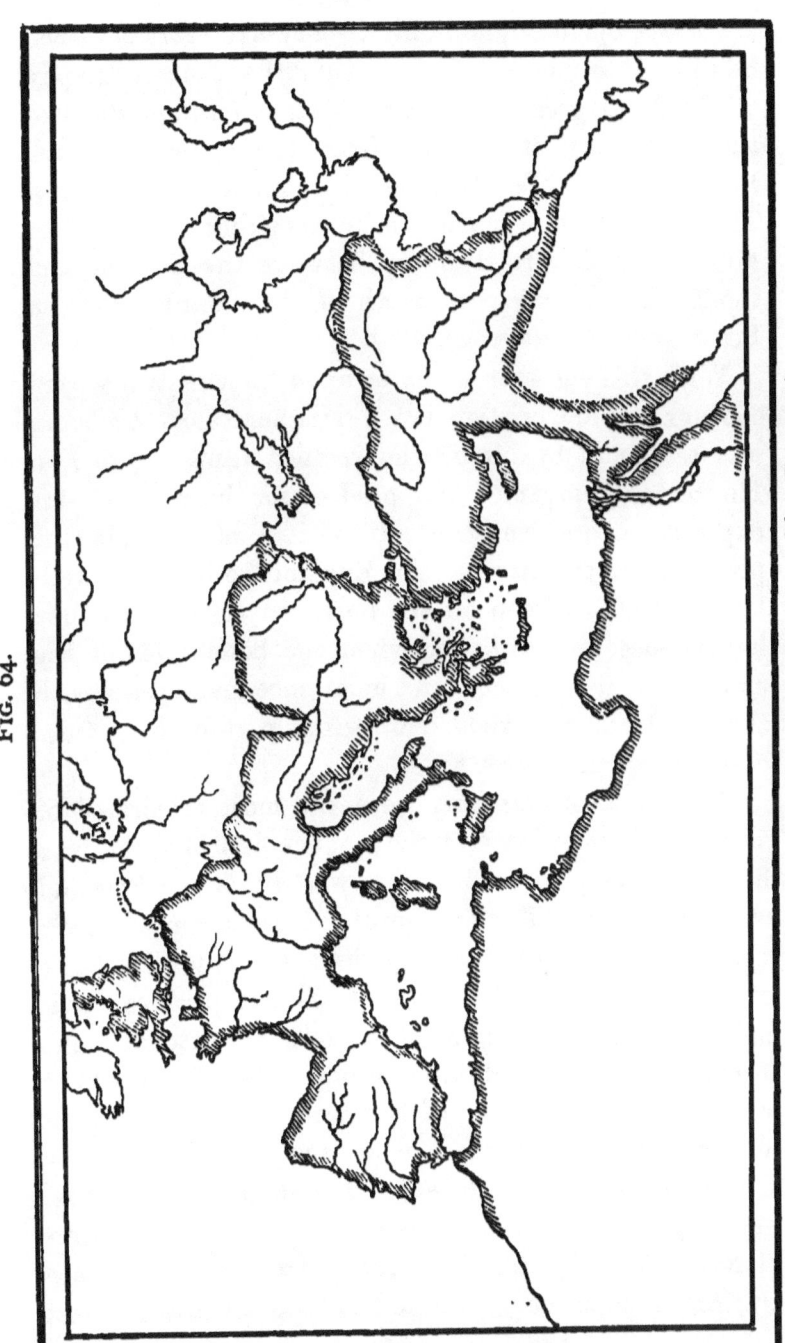

FIG. 64.

THE ROMAN EMPIRE UNDER TRAJAN. (*After Freeman.*)

He was a Spaniard, and only twenty-five years old when he came into power. In his reign the Roman Empire attained its greatest extent. Through conquests over the Parthians, it reached from the Caspian to the Persian Gulf, while Dacia was also subdued and annexed. Thus the empire never included either Bohemia or Moravia, or Austro-Hungary between the Danube and the Theiss, nor any part north of the Danube between Pesth and Donauwerthe.

With this emperor we enter upon the eventful second century of the Christian era. His successor, *Hadrian*, was a Roman by birth, who reigned from 117 to 138. He had accompanied his predecessor in most of his expeditions and subsequently visited almost all the provinces of the empire. In England he built the well-known wall to keep off the northern tribes. He had hardly succeeded to power when the boundaries of the empire began to recede, and he at once made peace with the Parthians, and yielded up great part of his predecessor's conquests over them.

The next two emperors were the most celebrated of all for their morality and piety. The first of these, who had been adopted by Hadrian, was *Antoninus Pius* and was the son of a Roman consul. He came into power 138 A.D., and reigned twenty-three years, leading a life of most exemplary benevolence and temperance. His only war was in England, wherein he extended the Roman dominion, building a second wall to keep out the unsubdued inhabitants of Scotland. He adopted his successor, *Marcus Aurelius*, who espoused his unworthy daughter Faustina. He succeeded to power 161 A.D., and reigned nineteen years with the esteem and admiration of his subjects. But many troubles took place during his supremacy, and he had to exert the greatest efforts to preserve the empire from further diminution. On his death, 180 A.D., a son for the first time succeeded his father. This was *Commodus*, who was a monster of vice and cruelty, and was murdered after a reign of twelve years. A worthy successor, *Pertinax*, was chosen, but was murdered in less than three months by the soldiery, who then sold the empire to *Didius Julianus*, who was beheaded after sixty-six days, when *Septimus Severus*, an African, having avenged Pertinax, reigned from 193 to 211. Thus his reign introduces us into the third century of our epoch. He was succeeded by both his sons, *Geta* and *Caracalla*, with such hatred and jealousy between them as a consequence, that the former was soon murdered. His successful rival, a monster who by his cruelty afflicted the whole empire, was in turn murdered 217 A.D.

After a brief attempt to hold power by the militarily elected *Macrinus*, he was succeeded by two Syrian youths, each a son of two sisters, daughters of Caracalla. The first, *Heliogabalus*, was a priest of the Sun, and in him Rome for the first time became subject to an Eastern sovereign. He was a wonderful example of effeminacy and vice, who, having been murdered by his guards 222 A.D., was succeeded by his worthy cousin, *Alexander Severus*, under whom the empire enjoyed unwonted happiness. He, in turn, was murdered A.D. 235, owing to a conspiracy of his successor *Maximin*, who, though born within the empire, was by blood a Goth.

From the time of the Antonines (Pius and Marcus Aurelius) the condition of the inhabitants of the Roman provinces was greatly ameliorated, and by the beginning of the third century all the inhabitants of the provinces, who were not slaves, had become Roman citizens. Thus the empire had become a monarchy—a homogeneous mass of people ruled by a chief who was more than a king, save for that title and the absence of a settled hereditary succession. As the city of Rome had thus gradually waned and lost all real sovereignty, so the power of the emperors had waxed and become established.

Alexander Severus was compelled to wage war with the Persians, who (in the year 226) revolted, and under their first king, Artaxerxes, occupied the territory of the Parthian kingdom.

Maximin, by his tyranny having excited universal hatred, was declared a public enemy by the Senate, which named two illustrious Romans, the *Gordians* (father and son), as emperors—only to lose their lives in little more than a month. Then *Maximus* and *Balbinus* were declared joint emperors, and Maximus, in the ensuing contest, was murdered with his son at Aquileia (238 A.D.), as were also his successful opponents in the same year. A third Gordian was then made emperor, but was assassinated and succeeded by *Philip*, an Arab raised to power by the soldiery, in the year 244. He celebrated with magnificence public games, and also the thousandth year of Rome's existence. He was defeated in a struggle with *Decius* who had been a senator, and became emperor 249 A.D.

Then the Goths (a people who, migrating from further north, occupied Dacia and crossing the Danube invaded the empire) descended into Illyria; when Decius, after successfully combating them, was drowned, 251 A.D. His successor, *Gallus*, was slain after a military revolt (in the year 253), and was succeeded by *Valerian*, who, in about his sixtieth year, was elected by the unanimous voice of the Roman world. He was a man of noble birth, learned, and of mild and unblemished manners, but who associated with his power his youthful but corrupt son *Gallienus*. Then the empire was attacked on all sides by the Teutonic tribes known as Franks and Alemanni, by the Goths and by the Persians. The Franks invaded Gaul and thence passed to Spain and onwards into Africa, while the Alemanni, crossing the Alps, penetrated almost to Rome itself, only, however, to be repulsed. The Goths, after dominating what is now Constantinople and part of Asia Minor, passed on (253 A.D.), ravaging Greece, plundering the Temple of Delphi, and threatening Italy. But Valerian felt bound to try and punish the hostility of the Persian king Sapor, and so crossed the Euphrates. He was defeated and taken prisoner (260 A.D.), and died in captivity. The power of Persia long endured, and was governed by the descendants of Artaxerxes (called the *Sassanidæ*) for four hundred years. Under the remaining sovereign, Gallienus, the whole empire became divided between various pretenders to the imperial power, who all met with violent deaths, while the universal confusion and disorder greatly diminished the population of the empire. In the year 268 Gallienus died and was succeeded by *Claudius*, who after gloriously routing the Goths at Nissa (in what is now Servia), died (270 A.D.), and was succeeded by *Aurelian* who, like Claudius, was an Illyrian. In a reign of but four years he recovered Gaul, with Britain and Spain, but had to relinquish Dacia to the Goths. Assassinated in the year 275, he was succeeded by *Tacitus*, who in little more than six months was again succeeded by an Illyrian named *Probus*. He in turn was killed by his soldiers in 282, after delivering Gaul, invading Germany, building a wall from the Rhine to the Danube, and recruiting the Roman army from the German tribes. Next came *Carus*, who delivered Illyria from the Barbarians, and having victoriously invaded Persia beyond the Tigris, died mysteriously in December 284. His two sons, *Carinus* and *Numerian*, succeeded him, whereof the latter died the following year. His unworthy brother alienated by his vices the hearts of the people, and *Diocletian*, commander of the Imperial Body Guard, was chosen emperor. Diocletian advanced to meet Carinus in Moesia, when the troops of the latter were entirely victorious, but in the moment of success he was stabbed in revenge for a private wrong (A.D. 285).

With the successful Emperor Diocletian began a new epoch of the Roman Empire, though he was but the son of parents who had been slaves, while his mother was a native of Dalmatia, which ultimately became his chosen residence. The changes he effected were greatly facilitated by the length of his reign, which lasted twenty years. He introduced an Oriental magnificence of royalty, assuming a diadem and introducing Persian ceremonial. His despotism was more concentrated than ever, though administered through an enormous bureaucracy. All provincials being formally acknowledged Roman citizens, the city of Rome lost still more in importance, while the royal residence was usually either in Milan or Nicomedia. Diocletian made important changes in the constitution of the empire, as to which the student is referred to histories of Rome. He abdicated the imperial dignity 303 A.D., and retired to Dalmatia, where he expired, having erected a magnificent palace the ruins of which still exist at Spalatro.

He had associated with him in power *Maximian, Constantius*, and *Galerius*, and, after his abdication, great disorder ensued. Constantius governed the province of Gaul and was favourable to Christianity. He died 306 A.D., and his son *Constantine* took his place. Maximian, who had abdicated with Diocletian, was replaced by his son *Maxentius*, who in the year 306 was declared emperor at Rome. Thereafter great confusion and dissension arose, there being six claimants for the empire, and Maximian re-entered the political arena. In 312 A.D. civil war broke out between Constantine and Maxentius. When near Rome, the former defeated his opponent, who was drowned and afterwards decapitated. Eleven years later, he defeated the last of his opponents. In the year 324 Constantine became sole sovereign of the entirely re-united empire. With Constantine two notable events took place: the first of these was that the seat of the empire was transferred from Rome to the newly built Byzantium, now Constantinople. The second event, which was of inexpressibly more importance, was that for the first time, in the person of Constantine, Christianity mounted the imperial throne, which thenceforth, save for the brief reign of Constantine's nephew, the Emperor Julian—from 361 to 363 A.D.—continued officially Christian till the extinction of the last shadow representing it was abolished, 1806 A.D.

Such being a brief sketch of landmarks in the civil and political history of Rome, it but remains to consider briefly the religion and philosophy of the Roman people. A knowledge of these, together with Roman political history down to Constantine, may enable the student to understand how the Roman civilisation, acting on races settled within the empire, has produced that European world which now exists.

The religion of the Romans was very different from that of Greece. Their gods—mostly survivals of the deities of the tribes whose union founded Rome— had hardly any legendary histories such as those which pertained to the divinities of Olympus, while they were served with greater devotion and more respect. Strong, however, as was the religious sentiment of the Romans, and important as everything was deemed which related to the worship of the gods, nevertheless this induced no tendency towards a theocracy. Religion was entirely subordinated to the State, and its priests had also civil functions, not as priests but as citizens: they were thorough laymen, and not animated by the sacerdotal spirit. Religion consisted of formal acts, and not in the acceptance of dogmas or in devout aspirations. The pious man was he who knew well how to honour the gods according to the laws of the State, and who presented himself in the temples properly dressed, in prescribed attitudes—the priest serving but as a master of the ceremonies, and to read the formulæ which the worshipper had to repeat.

It was a religion which accustomed the people to discipline and obedience, and was essentially much more moral than that of Greece. The Romans believed that it was through religion—the aid of the gods, especially Jupiter—that they had conquered the world, and Rome was considered by the Greeks, as well as by its own inhabitants, to be the most religious city in the world.

At an early period, however, many of the Greek and Roman gods became fused and confounded together; and other pagan religious influences from the East came afterwards to exercise a powerful influence on the beliefs and practices of the Italians.

Towards the close of the republic, a profound spirit of scepticism had, indeed, asserted itself in Rome; and Cicero and Cæsar (and the writings of Lucretius) may be taken as examples of the unbelieving spirit of Roman society in their day—though they upheld the external practices which were traditions of the State.

With the advent of Augustus, a renovation of Roman religion commenced. He sought its political support, rebuilt temples which had fallen into decay, introduced fresh objects of worship, and sought to institute moral reforms. He associated himself with the various priesthoods, and became *Pontifex Maximus* himself.

A practice arose of deifying deceased emperors, which, strange as it may seem to us, was but the continuation, or revival, of a practice of very ancient date in Egypt and elsewhere.* Ultimately the worship of a recently deceased emperor came to be a most conspicuous one in the Roman provinces, especially as a manifestation of loyalty to the State.

In the Augustan Age it became the fashion to admire the simplicity and piety of earlier days, and Horace, and above all, Virgil, were incited by the emperor to promote this sentiment. That well-known poem, the *Æneid*, of the last-named poet, was essentially a religious one, and has had great influence on religion down to our day.

The terrible times of Tiberius and Caligula were not likely to weaken the influence of religion, nor did the spirit of scepticism again raise its head at any period of the empire as it had under the republic.

As the Roman religion was devoid of dogma, so also was it devoid of any spirit of either proselytism or intolerance. Indeed the Romans for a time viewed with jealousy any worship by strangers of their Jupiter of the Capitol, from the same spirit which made them seek to wrest from their opponents the succour of their various gods, by themselves showing great respect and reverence

---

* See *ante*, p. 323.

to them. But although for a time the worship of foreign deities in Rome was forbidden, later on there was, as it were, an interchange of religious worship all over the Empire, and thus a sort of universal pagan church—a church of practices, not doctrines—preceded the diffusion of a universal Christian church.

So it came about that Isis and Serapis were worshipped at Rome in the second century after Christ.

The Oriental religions differed generally from that of Rome in their more sacerdotal and emotional character. Worshippers were invited to mourn over the death of an Adonis, to sympathise with a divine mother who sees the beauteous Athis expire in her arms, or to rejoice at the resurrection of a god such as Osiris. These practices, mixed with sanguinary rites and devotions of very doubtful morality, began to exercise great influence over the lower classes of Italy, and above all of Rome. But such beliefs and worships readily accommodated themselves to, and harmonised with, the State religion, as also did the widely diffused and very successful Oriental worship of Mithra.

During the period which elapsed between the days of the republic and the end of the second century, a tendency arose to recognise one God as supreme over all other gods, or even to revere one solely existing God; and this tendency grew and developed itself. Nevertheless the educated Romans of the period of Cicero, and for a time afterwards, were doubtful with respect to a future life. Under the empire, however, scepticism in this respect tended to diminish, while the great influence of that belief and the vivid fears it inspired, are attested by the number of associations formed (in which even slaves bore their part) for the purpose of providing their members with the requisite funeral rites and sepulture.

Jealous as was the State generally of all private associations, they were freely permitted for this one purpose, and for the feasts and sacrifices therewith connected.

After these few words we must turn from the consideration of Roman religion to that of *philosophy* at Rome. Roman philosophy, and Roman art also, were not products of the soil, but imported from Greece, which became, under the empire, the acknowledged model and leader in the arts and amenities of life. Despised in the earlier days of the republic, Hellenism became later a universal fashion. Thus, when Carneades* came to Italy, he was soon sent back to Greece by Cato the censor (the great-grandfather of the Cato before mentioned †), and philosophy did not become established in Roman society till the time of Cicero.

It was the school of Epicurus* and the rival school of the Stoics,‡ which successively prevailed at Rome, the former mainly in the later republic and the early days of the empire, but the latter afterwards. Stoicism may be said to have been the main and dominant philosophy, and it accorded well with all that was noblest and best in the tendencies of the Roman people. It was a philosophy which inculcated and promoted all charitable and benevolent movements; the kinder treatment of slaves, whose very lives were, at first, by law, entirely at the mercy of their masters, and who formed so very large a part of the entire population.

The most illustrious of the Roman Stoics was Seneca, who began to write under Caligula, was exiled by Claudius, then recalled and made a tutor of Nero, who ultimately sentenced him to death on a charge of treason. The

---

* See *ante*, p. 330.   † See *ante*, p. 345.
‡ See *ante*, *loc. cit.*

teaching of Seneca was of such a refined and elevated morality, that it has been mistakenly attributed to Christian influence. That Stoicism penetrated and prevailed throughout Roman society is shown by the fact that one of its most celebrated teachers was the slave Epictetus, while it ascended the throne with the Antonines, being encouraged by Antoninus Pius and actually professed and taught by Marcus Aurelius.

By the end of the second century many sorts of philanthropic and charitable efforts had been instituted in Rome, and such tendencies increased, at the same time with devoutness in religion, solicitude about a future life, the sacerdotal spirit, and a tendency towards monotheism.

But a new departure in philosophy became developed in Egypt from the combined influences of Hellenic culture and Jewish religion. This was the school of the *Neo-Platonists* of Alexandria which, beginning in the days of Augustus, greatly influenced the Roman world. Its effects indeed are widely diffused even at the present day. Alexandria had developed enormous commercial industry and, with its consequent wealth, had become a meeting-place for various races and civilisations, and a great intellectual centre, under dominant Hellenic influence. The first renowned member of the new Alexandrian school was the Hellenic Jew Philo, who was born about 25 B.C. His philosophy was essentially theological. He taught that the only true existence, in the highest sense, is God, dwelling apart and acting on the world through an intermediate being, the "Word" (in Greek *Logos*), dwelling with God as his wisdom (in Greek *Sophia*), and emanating into numerous subordinate spiritual agencies. The "Word" was not eternal but generated by God in a peculiar manner, and it was this His "first begotten" who had created the world. As to what it was desirable to do, Philo taught that logic and reasoning are of relatively little value, and that the highest philosophy consisted in the acquirement of a direct intuition* of God, to be attained as a reward for complete self-renunciation and resignation to divine influence.

Thus the Neo-Platonic school was essentially mystical, and renounced all appeals to reason as the one only source of the highest knowledge. Philo visited Rome in 40 A.D., and doubtless there exercised a direct influence. But the school of Alexandria accompanied the rise, and combated the spread, of Christianity.

The religion destined to play so great a part in transforming the world naturally appeared at first to the Greeks and Romans as an obscure sect of the Jews. The latter were early regarded as an impious race which despised the gods, and would not ever worship the emperors, for which refusal they suffered cruel persecutions. The inhabitants of Antioch burnt alive many who would not abjure their creed, and fifty thousand Jews are said to have been massacred at Alexandria.

The Pagans could not understand the obstinacy of Jews and Christians in refusing to worship Roman gods, whose clients were so willing to venerate what was adored by either Christians or Jews. Romans were willing to recognise in Jehovah their own Jupiter, and Alexander Severus † erected an image of Christ amongst the deities of his private chapel. But the refusal of Christians to worship the gods in any way, and their repeated declarations that the gods of the empire were but idols to be execrated, drew on them the persecutions and sufferings they had to endure, not only under

---

* As to this word, see *ante*, p. 261.   † See *ante*, p. 351.

monsters like Nero, but under such virtuous sovereigns as Antoninus Pius and Marcus Aurelius. In 202 A.D., Severus instituted a persecution which lasted about ten years. A short but very severe and universal persecution also took place under Decius,[*] which was renewed by Valerian, 257 A.D. In 303, Diocletian ordered all the sacred buildings and books of the Christians to be destroyed, and his persecution was one of the most severe of all; but it was the last.

The exceedingly rapid spread of Christianity through the empire is largely to be accounted for by the preparation unconsciously made for it in pious minds by that fermentation of religion and philosophy at Rome, which took place during the first two centuries of our era, as has above been briefly indicated. The changes which thus prepared the way for Christianity were: (1) the revival of Roman religion under Augustus; (2) its progress up to and beyond the time of the Antonines; (3) the inculcation of benevolence and philanthropy; (4) the anxiety as to the state of the soul after death, and the fears of such infernal punishments as were depicted by Virgil; (5) the increasing tendency towards monotheism; and (6) the sacerdotalism introduced with the advent of Pagan religions from the East.

To the wants thus indicated, Christianity marvellously responded, and the good tendencies previously developed were by it greatly intensified. The old system, without Christianity, could never have adequately responded to the needs and aspirations of the men of those days, and for the following reasons: (1) however strong or widespread might have been the tendency amongst the Pagans to attain to a conception of the unity of God, it was impossible for them, while maintaining their polytheistic

---

[*] See *ante*, p. 352.

worship, to arrive at a distinct, well-defined, and general consent concerning it; (2) the moral advance made under Paganism was considerable, but so long as the legends of the gods were tolerated, and certain of their rites practised, any approach to a perfect system of morality was unattainable; (3) popular devotion, however stimulated by introduction from the East, remained essentially material and devoid of exalted aspirations; (4) philosophy, though appealing to the cultured few, made no sufficient efforts to attain the popular ear and reach the masses; (5) it was impossible to avoid a profound distinction between the ideas and convictions which actuated the well-disposed men of education, and those which acted upon the bulk of the population; (6) need was felt for a religion consisting not of external forms, but of distinct and definite dogmas, fundamentally identical in their claim upon the assent and obedience of rich and poor, and accompanied by distinct unequivocal precepts. The Christian religion, so uncompromising and distinct in its commands and dogmas, coupled a lofty morality with the foundation of its faith; its enunciation of the unity of God was plain and unequivocal, and its theology was free from all admixture with gross and criminal fables. It appealed both to the cultured and the uneducated, and it commanded obedience of the will, while it incited the emotions to aspire after the loftiest conceivable ideals. Its spread was greatly aided by that freedom accorded to associations for performing funeral rites, before noticed. In this character Christians were easily able to assemble for worship and mutual edification, and we may see abundant evidnces of this in the Catacombs of Rome to-day. Christianity had already spread far and wide in the time of Decius. It became socially predominant under Constantine by whose authority the well-known Council of Nice was convoked. Under his son Constantius, Paganism began to be persecuted, and Julian (the son of a brother of Constantine), in his fruitless attempt to revive Paganism, was rather seeking to oppose to Christianity a new Hellenic religion of his own devising, than trying really to restore the old worship of Rome and Italy. The school of Alexandria, though it opposed, really helped to promote and develop, the theology of the Christian Church; and ultimately, in the time of the Emperor Theodosius (367 to 375 A.D.), all public profession of the religion of Pagan Rome finally ceased.

With this brief and elementary sketch of the history of the rise and development of the Christian empire we must here content ourselves, referring the student to the various historical works wherein he will find depicted the struggles which eventuated in (1) the destruction of the western half of the empire; (2) the establishment of the existing European nations; (3) the rise and gradual extension of the recognised prominence and and power of the Bishop of Rome (which followed so naturally the traditions of the city of Rome, as having possessed in an eminent degree the genius of legislation, orderly rule, and universal dominion); (4) the temporary eclipse of antecedent culture and civilisation; (5) the gradual awakening of the spirit of inquiry; (6) the discovery of the previously unknown half of the globe; (7) the schools of philosophy which from time to time arose up to our own day; (8) the rapid advance of physical science; and (9) the rise of the conception of human progress. These can be but mentioned here, our intention being to present our readers with no more than an introduction to such elementary historical knowledge as may serve to incite them to pursue the great science of human history.

CHAPTER X

SCIENCE

The beginner who has attentively read the preceding chapters of this little work will have been introduced to the elements of the *physical sciences* and also to those of *psychology* * and of *logic*.

We saw at the outset † that there is one property common to all those things we have successively passed in review—namely, the property of *number*. But there is another quality which is also common to every well-founded perception we have gained, whatever has been the nature of the objects considered. That quality, or property, is *truth*. ‡

It is a matter of course that no scientific conclusion, or doctrine, can be worth anything if it is not "true," as also that every scientific inquiry is necessarily an inquiry after "truth"—after truths of one kind or another.

Now the tests of truth are different in different lines of investigation. In mechanics and the study of physical forces we may often avail ourselves largely of experiment, as also in the study of the living world. In *Palæontology*, § however, we can but make use of

---

* See *ante*, pp. 252 and 268.   † See *ante*, p. 5.
‡ See *ante*, p. 260.   § See *ante*, p. 244.

observation and of inference, and inference plays a yet larger part in many *historical* investigations.

But we may occupy ourselves not with what may be the truth in this or that scientific inquiry, but about the different kinds of truth attainable in different inquiries. The sort of truth to be gained by the study of the geometry of Euclid is evidently different in nature from that afforded by investigations as to animal physiology.* This leads us to the question of the different degrees of truth which different inquiries may be able to afford us, and so we may pass on from considering what are the tests to be applied, and their value, in different branches of knowledge, to the question whether there is any one test we can apply to the ultimate and fundamental propositions of every branch of knowledge; and if so, what that one test may be?

Thus the various sciences have each their own fundamental truths which are vouched for and shown to be valid by their own proper tests. But we should endeavour to see whether there are any truths which underlie all sciences, and if so what they are, and what, if anything, vouches for them and establishes their validity.

In this chapter, then, we propose to introduce the student to a knowledge of the elements of "Science *par excellence*" and to consider what, if any, test can be found, not for truths of this or that order, but for all truths of whatsoever order.

In this consideration we shall have to make use of reasoning, as we have again and again made use of it, and no science can be properly followed up without the aid of reasoning.

Now we all know that science has greatly advanced

---

* See *ante*, p. 187.

since the beginning of this century. Many things are now known with absolute certainty which before it were unknown. No one will probably deny that science is still advancing. Nevertheless such advance would be impossible if we could not, by observations and reasonings, become so certain with respect to some facts, as to be able to make them starting-points for fresh observations and reasonings as to other facts. It is impossible to deny that we may repose with absolute confidence and entire certainty upon a variety of such newly ascertained facts; *e.g.*, that as to the mode of reproduction in ferns.*

The laws of the reasoning process and the conditions of its validity, have been explained in an elementary manner in the chapter on logic. But however excellent and admirable our process of reasoning may be, the process must stop somewhere. For in order to prove anything by reasoning, we must show that it necessarily follows, as a consequence from antecedent truths, which therefore must be deemed yet more indisputable. But this process cannot go on for ever. We cannot prove everything. However long our arguments may be, we must at last come to statements which have to be taken for granted, otherwise we should have to reason for ever, which is, in effect, to affirm that nothing at all can ever be proved. If we could not know some things without their being proved, we could never reason validly at all. So again with respect to the validity of the reasoning process itself. When the laws of logic have been duly complied with, the validity of that process is a truth which we can see to be true by studying it, and which is also implied in the fact that science progresses. If

---

* See *ante*, p. 205.

we had to prove the validity of that process, such a proof would require a second proof, that again would require a third, that third a fourth, and so on for ever. In other words, there could be no such thing as proof at all.

Now no one who argues, or who listens to, or reads (with any serious intention) the arguments of others, can, without stultifying himself, profess to think that no process of reasoning is valid. If the truth of no mode of reasoning is certain, if we cannot draw any certain inferences whatever, then all arguments must be useless, and to proffer or consider them must be alike vain. And not only must all reasoning addressed to others be thus vain, but the silent reasonings of solitary thought must also be vain. This is equivalent to declaring all men to be idiots. But a man who honestly considers all other men to be mad, is generally, and not unreasonably, deemed to be somewhat intellectually deficient himself. Thus the validity of the process of reasoning can but depend upon its own evidence. It depends on the clear and manifest evidence it possesses for any one who will consider it and who understands the meaning of the idea "therefore"—when plainly showing how one truth results from another—how, *e.g.*, the mortality of Socrates is involved in the fact that mortality is common to the whole human race.

Thus the validity of the *process* of reasoning is evident in itself; but the truth of any particular inference depends, and must depend, upon certain facts which are known to us independently of and without reasoning. We could not, for example, conclude that the fact of some animal having a spinal column,* proved it to be a

---

* See *ante*, p. 234.

"vertebrate," if we did not know that all vertebrate animals have a spinal column.

If we had no certain knowledge of any fact, all our processes of reasoning would, as it were, remained suspended in the air and have no relation with any real thing whatever. Now our knowledge of facts may vary greatly. Some men know an enormously greater number of them than others do. There is, however, one fact which everybody knows with the most absolute certainty, and that is the fact that he himself exists.* This knowledge lies at the foundation of our whole intellectual life.†

The supreme and absolute certainty which each of us may have with respect to this fundamental fact would seem to be indisputable; and yet there are some persons who dispute it and affirm that though we may, and do, have absolute certainty about any state of feeling present at the moment, no continuous or substantial self-existence can certainly be known to us.

Now, we have already affirmed it ‡ absurd to suppose our consciousness is made up of an aggregate of states, and that we have no continuous intellectual activity. It is now time to show how the delusion that we cannot be supremely certain of our own continuous existence has arisen. Our primary, direct consciousness at any moment, is neither a consciousness of a state (of a "feeling") nor of our continuous existence, but a perception and consciousness of our doing something (*e.g.*, reading this book) or having something done to us (*e.g.*, being supported in a chair). We have, indeed, all the time some vague perception of "self-existence" and

---

* See *ante*, pp. 258 and 270.   † See *ante*, p. 258.
‡ See *ante, loc. cit.*

some states of "feeling," but what we perceive primarily, directly, and immediately is neither the "feeling" nor the "self-existence," but some concrete actual doing, being, or suffering then and there experienced.

Any one of us can, if he pleases, turn back his mind upon itself, and either say to himself, "I have the feelings which accompany reading this book," or "it is I who have these feelings." But neither of these two acts is a primary act of the mind. No one in beginning to think, adverts either to his "present feelings" or to his "continuous existence." No one begins by perceiving his perception a bit more than he begins by expressly adverting to the fact that it is he himself who perceives it. But to become aware that one has any definite feeling, is an act of reflection which is at least as secondary and posterior as it is to become aware of the "self" that has the feeling. Indeed a more laborious effort is needed to recognise explicitly the implicit "feeling," when we know we are doing anything, than to bring before the mind *explicitly* * the implicit perception of our "persistent existence." Men continually and promptly advert to the fact that actions and sufferings are *their own*, but do not by any means so continually and promptly advert to the fact that the feelings they experience are "*existing feelings.*"

One of the greatest and most fundamental of all errors, is the mistake of supposing we can know our "states of feeling" more certainly and directly than we can know the continuously existing self which has those feelings.

And how is the truth of this perception known to us? What is the test whereby we know it to be true?

---

* See *ante*, p. 255.

We know it to be true because we immediately perceive it. It is an evident intuition,* and its test is its own luminous self-evidence—its evidence in and by itself. We can never *prove* to ourselves our own existence, it is a self-evident fact which our intellect directly perceives. Closely connected therewith is our power of memory.† Our continuous existence is therein implied, and the validity of our faculty of memory is also implied in every scientific experiment we perform. It is plain that it would be impossible for us to be certain about any careful observation or any experiment, if we could not place confidence in our memory being able to vouch for the fact that we had observed certain phenomena and what they were.

By asserting the general trustworthiness of our faculty of memory we do not, of course, mean to deny that mistakes are often made. Nevertheless we are all of us certain as to *some* past events. Every reader of this book, for example, is absolutely certain that he was doing something else before he began to read it.

Memory gives us as much certainty concerning *some* portions of the past as we can have with respect to *some* portions of the present. If we could not trust our faculty of memory, all science would be, for us, a mere dream. But the veracity of that faculty is a self-evident truth. It can never be proved. There can be no such thing as *proof* of it, because we cannot argue at all unless we already trust it.

There is another important fact with respect to memory. It not only tells about our own past, but about various facts with respect to other persons and

---

* See *ante*, p. 261. † See *ante*, p. 259.

to things external to us, of which we had knowledge at some past time, as we shall see later on.*

But if we were provided with nothing more than absolutely certain evidence as to a number of facts, together with a perception of the necessary validity of the reasoning process, we could make no use of such knowledge unless we were also provided with some absolutely certain general principles. Without them we could have no sure basis for the premisses of our syllogisms. Thus, for example, we could not conclude that Socrates was mortal because all men are so, save for the general principle: "whatever is mortal cannot at the same time be immortal."

Such general principles are no less indispensable for physical science than for syllogistic reasonings.

For every experiment carefully performed implies a conviction on the part of him who performs it that such general principle can be relied on with certainty. Let us suppose that the experiment of cutting off an eft's leg has been performed in order to see whether a fresh leg will grow, and let us further suppose that a fresh leg has grown; this experiment will have demonstrated that such a thing is possible, because, in fact, it has actually occurred. But that certainty implies a prior and much more important truth. It implies the truth that if the eft has come to have four legs once more, it cannot at the very same time have still only three legs. If we reflect again on this apparently trivial proposition, we shall see that it depends on a still more fundamental truth which our reason recognises—the truth, namely, that "*nothing can at the same time both be and not be*"—which truth is known as "the law of *contradiction*."

---

* See *post*, pp. 385 and 386.

This is another truth which carries with it its own evidence and is incapable of proof. That such is the case a little reflection will show. We act upon it constantly in daily life without adverting to it. We know that if we have spent all our money we can have none left in our pockets, and we see plainly that any man who really doubted whether, if he had lost one eye, he might not at the same time still retain his original two eyes, or who really doubted whether, if his legs were cut off, they could not at the same time remain on, would have a disordered mind. The simplest rustic knows that if his wages have been paid to him, they are no longer owing, and that if he has put his cart-horse in the stable, it is no longer between the shafts. Yet the "law of contradiction" is but the summing-up, in an abstract formula, of a multitude of particular instances of this kind, as to each one of which no doubt is, or can be, for a moment seriously entertained by any sane mind. If we were really to doubt about the law of contradiction we should fall into mere folly. For if anything can at the same time both be and not be, then nothing can be *true* without its being possible for it also to be *untrue*, and this amounts to a paralysis of the intellect.

But if we consider any distant period—such as the days of Julius Cæsar—or any distant spot, such as the surface of the moon or even the star Sirius, we can see clearly that the same law must necessarily also apply then and there. Such a truth is therefore distinguished as an absolute, necessary, and universal truth or principle— one applicable to all times and all places whatsoever.

Another truth of a similar character, but less universal application, is an axiom we before [*] noted, namely,

---

[*] See *ante*, p. 34.

that "things which are equal to the same thing are equal to each other." This is an abstract, universal, and necessary principle, which our reason can apprehend to be such, although it may never before have been adverted to, save as implied in ordinary facts of experience. If, for example, a man has found that two pieces of wood are each of them equal in length to a third piece of wood which is a yard measure, he will at once be certain that the first two pieces are of equal length — namely, each a yard long. From a variety of instances of equalities of very different kinds, he will be led to appreciate the force of the principle just quoted, when his attention has once been directed to it. It will then be evident to his mind that this equality between the equals of a third thing must positively always and everywhere exist. In our perception of the truth of this principle—this law—some other very fundamental principles are necessarily involved, as will become obvious if we turn the mind inwards upon itself. Thus it is obvious that this law, as it concerns equality generally, must concern every kind of equality—equality not only between "quantities," but between "qualities" and "relations" also. Two daughters and a son of the same mother are all equally her children, and if she feels the same amount of love for each girl as she does for her boy, then the love felt for one girl will be equal to that entertained for the other.

Things which agree in quantity, quality, and relation are in so far alike. Yet they cannot be thought of as being "alike," unless they are at the same time recognised by the mind as being existing things which are *distinct*. Thus in the above axiom we have involved the ideas "distinctness," "similarity," and "existence," as we before saw * to be the case with respect to our idea of "number."

This axiom about equality, though it can be illustrated by any number of instances, can never be proved by reasoning. It is a self-evident truth which reposes on its own evidence—as do the other axioms which are characterised as "evident" in the second chapter of this work.

Self-evidence, then, is our ultimate ground for assenting to such axioms, as well as to necessary principles and to the fact of our own existence.

We have already pointed out † how universal is the desire of mankind to know the *causes* of circumstances and events. To know this is, as before said, the aim and object of the highest form of science. Particular sciences may be devoted to ascertaining that certain things are, and the circumstances of their being—their successions and co-existences—but no such knowledge of mere phenomena will suffice to constitute *science itself*. Such science is but another term for *philosophy*—the science of sciences—which is what we are concerned with in this chapter.

If we examine our minds as to what our idea of "cause" is when that conception is called forth, we shall see that it stands in close relation to our perception and idea of "change."

When some change occurs, or when anything strikes us as being a new thing, we spontaneously look out for its cause. The truth concerning causation, which our mind recognises as being necessary and self-evident, is the principle that "*every new existence is due to some cause*"—and this is quite in harmony with that spon-

---

* See *ante*, p. 254.   † See *ante*, p. 268.

taneous habit of mankind before mentioned. It is also a truth, the self-evidence of which a little reflection will make clear. Thus, a thing which is new cannot have caused itself, because it could never have acted before it came into existence. It must, therefore, have been brought into being by something else. But every change which takes place in a thing already existing is also, to a certain extent, a new existence, for it is a new state or mode of existence. It must, therefore, either be due to some antecedent mode of existence or to something distinct from it. Thus, if a door which was open is now no longer open, it must have been acted on by something else—a current of air, or what not. If a cat is now awake which was asleep, this must be due either to something external which has awakened it, or to some vital action of its own frame, which has aroused it from its dormant condition.

Again, all and every object made known to us by our senses is seen to be necessarily the product of some cause or causes external to itself. This is, of course, most manifestly the case with every product of human art; but every stone which we tread upon, or every patch of sand or mud, must have come to be as it is through the action of antecedent causes and conditions which have made it to be as it is and not otherwise. Not only the more or less complex structure of any solid body, but its size, position, divisibility and its existence at the time and manner in which it does exist, are all due to antecedent actions of other things which have determined its present conditions of existence. Even a portion of matter, which, so far as we know, is not made up of other material substances—such, e.g., as a diamond or a piece of gold—demands a cause for its relation to things around it and for its own size and internal conditions. The two latter circumstances would demand a cause for their being such as they might happen to be, even if such a body had existed eternally in an eternal universe. Everything then which can be known not to have a sufficient cause of its existence within itself must be due to some cause or causes external to it. Only something which is absolutely simple, indivisible, and eternal, can escape from this law of universal causation. This perception of the need of a cause is not the result of a mere *negative* impotence to conceive otherwise, but is a *positive* perception, as the reader may soon see for himself if he will examine into his own mind. Let him think of a stone peculiarly shaped, and then see whether his intellect tells him plainly and positively that some unknown cause or other must have determined its peculiar shape, or whether his mind is a mere blank with respect to it.

The idea of a cause is closely connected with the conception of "power" or "force"—ideas generated doubtless by our consciousness of our own power to think or act and by our active exercise of those powers. But the idea of "power" is a primary ultimate idea which cannot be resolved into other more fundamental or elementary conceptions. If the reader doubts this, let him try whether he can himself so resolve it. We never of course see the "power" itself which is thus exercised. If we strike a billiard ball with a cue we see the cue and the ball, the blow and its effects. But we can never see the influence itself which the cue communicates; for it is invisible as well as intangible. But though our *senses* cannot perceive it, our *intellect* can, and there is one instance, at least, wherein the inflow and action of causation is distinctly perceptible to us. This is our perception of the inflow, of the influence, of motives upon our will. When we resolve, from some motive to perform an act, we are conscious not merely of the existence of that antecedent state of things which is named "a motive," and of that consequent which is our "resolve," but of the motive also, as something impelling us. We know and feel that it is active and exerting an influence upon us; that is, emits, as it were, a force stirring our will. We have also an experience of the force of causation when anything resists our will. In the latter case the influence is antagonistic to an act of will already formed; in the former case the influence excites towards the formation of such an act of will.

Thus the "law of causation" is a truth borne in upon us by its own evidence, not only spontaneously in each instance of it which comes under our notice, but on reflection also; and the more we reflect, the more we see the evident truth and universal necessity of the law that "*every new existence is due to some cause*," which law is as certain as the law of contradiction itself. For if that which has as yet no existence could nevertheless be a cause, then it would no longer be the case that nothing can at the same time both be and not be. But as to the nature of the cause acting in any case, our power of intellectual intuition tells us little save that it must, in each case, be adequate to produce the effects to account for which it is invoked. No child with a toy hammer could level the great pyramid of Egypt; no ignorant peasant could translate a play of Æschylus; and no being devoid of intellect could perform a truly virtuous action, or make known an ethical idea—for moral perception is but one form of intellectual activity. That a cause must be adequate in order that a given effect may be produced is, an absolute, universal, and necessary truth, no less than is the law of causation itself.

It is indeed a very wonderful thing that we should be able to know that any truths are "necessary" and "universal"; but if we think the matter all round and consider some of our other faculties, it will not then appear to be so exceptionally wonderful. For *all* our knowledge is wonderful when we consider it deeply. It is wonderful that objects about us, acting on our organs of sense, should give us sensations such as those of musical tones, sweetness, blueness, or what not. It is wonderful, again, that by means of combinations of sensations actually felt and others remembered, we should be able to perceive surrounding objects.* It is also wonderful that we should be able thus to know not only our present circumstances, but also something of our own past. In the same way it is wonderful we perceive that if a thing "is" it cannot at the same time "not be." But nevertheless we do most certainly perceive that such is the case. We know some things and we know that we know them,† and amongst them we know this necessary truth and also other necessary truths. But *how* we know them (or *how* we know anything) is a problem to attempt to solve which is hopeless.

The mystery of our power of *intellectual perception* is parallel with that which attends our power of feeling. We feel things savoury or odorous, or brilliant or melodious, as the case may be; and by dissections and the microscope we may investigate the structure of the organs of sense and of the brain. But *how* those anatomical conditions can give rise to the feelings themselves, is a mystery which defies our utmost efforts to penetrate. No one knows *how* knowledge is possible, any more than *how* sensation or *how* life is possible, or *how* solid objects can have length,

---

* See *ante*, p. 254.   † See *ante*, p. 270.

breadth, and thickness. But ignorance *how* a thing has come to be, does not in the least deprive us of our certainty of the fact *that* things are—that we *do* know and feel and live, and that our body has the above-mentioned dimensions—*i.e.*, that it is "*extended.*" That we have a power of feeling is evident to us by our senses. That we exist, and know universal and necessary truths, is plain because our existence and such truths, have their evidence in themselves and need no proof—the ground on which we believe them is that they are *self-evident* to us.

But the student may ask whether there cannot be a better criterion of truth than "self-evidence." Let us then see whether or not it is possible to have a better. Now, whatever the supreme and ultimate test of any alleged truth may be, it must either reside in that truth itself—so making it luminously self-evident—or in something else, something external to it. But the value of any external criterion could only be appreciated by us through our perception of its cogency. If we suddenly saw the law of contradiction written on a cloud, or on the surface of the sun, such a fact would not add to its cogency. In the first place, various investigations might thereby be set on foot, such as the sanity of a witness, or the probability of a common simultaneous hallucination, and so on. But no one of such investigations could come to any conclusion except the principle of contradiction were first absolutely accepted. Thus if a universal or necessary truth was vouched for by some external criterion, the cogency of which existed in that criterion itself, we should then only have self-evidence after all, and "self-evidence" at second hand, as regards the ultimate truth for which such external criterion was to vouch. If, on the contrary, the external criterion has to be vouched for by something again external to it, we should have to go on in that way for ever, or else rest at last in something which was self-evident. It will be plain, on reflection, that nothing external—no common consent of mankind, no amount of testimony, no experimentation or evidence of the senses—could ever take the place of an ultimate criterion of truth, since some judgment of our own mind must always decide for us with respect to the existence and value of such criteria. Self-evidence, then, is really the only possible ultimate test of truth, and must be accepted under pain of complete intellectual paralysis. It is incapable of demonstration, since it depends on nothing else. It is continually made use of as a matter of course and without reflction, and is relied on constantly by every one who reasons. We have said that self-evidence must be accepted under pain of complete intellectual paralysis, because if it be not accepted, we are logically reduced to absolute scepticism—that is, to utter folly.

It is necessary that the student should recognise the fact that we are all of us certain of something. Although sincere inquiry, and therefore doubt, are not only legitimate but necessary in science, nevertheless reason shows us that doubt must have its limits. There is such a thing as *legitimate certainty*, otherwise there would be no certain science of any kind. But if there is such a thing as legitimate certainty, then to doubt about it must be *illegitimate doubt*. But, as we all know, credulity is common to children and to weak-minded, or ignorant, men and women; therefore extreme scepticism may, at first sight, seem to be an exceptionally intellectual state of mind. Nevertheless a little reflection will show that it is, in fact, an exceptionally foolish state of mind; for nothing can be more foolish than to contradict oneself; and that is what extreme sceptics are compelled, by pitiless logic, to do. We have met with a notable example of this folly in the case of Pyrrho* and his disciples the Pyrrhonists, who refused even to affirm the truth of their own system. If any one should venture to say "Nothing is certain," he would necessarily contradict himself, for he would thereby, in the very same breath say, or at least imply, that "something is certain," since he implies that we may be *certain nothing is certain*. He says, or implies, therefore, something which, if true, absolutely contradicts what he has declared to be true. But a man who affirms what the system he professes to adopt forbids him to affirm, and who declares that he believes what he also declares to be unbelievable, can hardly complain if he is called foolish. If he denies that he affirms anything, or even that he doubts about everything, then, since he affirms nothing, we may disregard his avowed folly, and the utter impotence he confesses to.

Let us see a little further how self-refuting sceptical modes of thought are. Suppose a man were to say: "I cannot be sure of anything, because I cannot be certain that my faculties are not always fallacious," or, "I cannot be sure of anything, because, for all I know, I may be the plaything of a demon who amuses himself by constantly deceiving me"—in both these cases such a man would simply contradict himself. For how can he know that "constantly fallacious faculties," or, "a demon deceiving in all things," will necessarily deprive him of certainty? Obviously he can only know this, because he sees the necessary truth: "*it is impossible to*

---

* See *ante*, p. 329.

arrive at conclusions which are certain, by means of what is uncertain or false." But if he knows *that* truth—if he is certain of that universal principle—he must know that his faculties are not *always* fallacious, and that his demon is impotent to deceive him in *everything*—since he can apprehend the certainty of the necessary truth last stated. *Universal* doubt, then, is simply an absurdity. It is scepticism run mad, and no system can be true, and no reasoning can be valid, of which the inevitable result is scepticism of such a nature.

Let us now see what would be the consequences of uncertainty as to the validity of the reasoning process and our mental faculties generally, of memory, of our knowledge of our own existence, or of universal necessary principles, or of the law of contradiction.

As to reasoning,* and the trustworthiness of our mental faculties generally, if any man is convinced that thoughts are worthless tools, he can only have arrived at that conclusion by using the very tools he professes to consider worthless. What, then, ought his conclusions to be worth even in his own eyes? It is simply impossible by reason to get behind or beyond conscious thought, and our thoughts are, and must be, our only means of investigating fundamental problems.† If a man professes to affirm that his faculties are untrustworthy, we of course are thereby debarred from bringing home conviction to his mind. But it is no less impossible for him to defend his position. Once let him attempt sincerely to do so, and he thereby plainly shows that he really has confidence in reason and in language, however much he may verbally deny that he has it. It

---

* For the consequences of scepticism as to this, see pp. 368 and 369. † See *ante*, p. 270.

is clear that we all do know and are certain about something, if only that we are inquiring about the certainty of knowledge. Utter folly, then, logically results from doubt as to the validity of reasoning.

Evidently, mental paralysis must also result from any real doubt about our own existence, and our perception of this existence involves the validity of our faculty of memory, which is implied in this way as well as in every scientific experiment we perform. For we cannot obviously have a reflex perception* either of our feelings or our self-existence, without trusting our memory as to the past.

Now there are two technical terms with which the student who desires to be introduced to the elements of science must make himself familiar.

They may be represented to a certain extent by the two familiar words "*facts*" and "*feelings*." The two technical terms which correspond with them respectively are: (1) things which are *objective*, and (2) things which are *subjective*.

Every "feeling" or "state of consciousness" present to the mind of whoever is the subject of it, is spoken of as being "subjective," and the whole of such experience, taken together, constitutes the sphere of "*subjectivity*."

On the contrary, everything whatever which exists externally to our present "consciousness" or "feelings," is spoken of as being "objective," and all that is thus external to the mind is the region of *objectivity;* it is the region of *facts* as considered apart from *feelings*. But the reader must understand that the feelings of one man are "objective facts" to any other man; only a man's own present thoughts and feelings are to him "subjective."

---

* See *ante*, p. 259.

Now memory informs us with absolute certainty as to some events of our own past history. But such events are beyond our present experience, therefore they have a truth which extends back beyond any present feeling we have. They are realities existing in addition to our present feelings, and they are therefore "objective." Thus memory, insomuch as it reveals to us part of our own past, reveals to us what is "objective," and so actually introduces us into the *realm of objectivity*, shows us more or less of "objective" truth, and carries us into a real world which is beyond the range of our present feelings. This power which memory has of thus lifting us, as it were, out of our present selves and showing us such facts, is certainly a very wonderful power; and yet its existence must be admitted, at the price of otherwise falling into a most absurd and idiotic scepticism which would deprive us of all power, not only of reasoning with others, but of having any rational perceptions ourselves. The moment lasts but a moment. If we could not know with certainty more than the passing moment, we could not really know even that. We could know *nothing*.

As to necessary truths, if we were to doubt about the "law of contradiction," we could again be certain of nothing. If we would rise from such intellectual paralysis we must accept that dictum as it presents itself to our minds; and that dictum declares that nothing can both be and not be; *e.g.*, that no creature, anywhere and anywhen, can, at the same time, be both bisected and entire. As to the "law of causation," its doubt or denial would annihilate physical science, and also, as we have seen,* would involve a denial of the law of

---

* See *ante*, p. 379.

contradiction, and so reduce us to the idiocy of absolute intellectual paralysis.

But we all know that we have a body as well as a mind, and that the body has the dimensions before mentioned,* *i.e*, is "extended." Similarly we perceive that other bodies are extended also, and we do this not by any process of inference† from sensations arising in us through the influence of objects about us, but by a direct intuition which such sensations occasion. It is not a synthesis of such sensations; for they continue to exist separately and can be recognised by the mind as doing so and standing as it were beside the direct perception they have occasioned which persists during the recognition of the sensations which have occasioned it. No mistake can be greater than that of thinking a perception is made up of the feelings which minister to it. Two marbles are *objectively* two, and can be so perceived, whatever the *subjective* impressions which have occasioned them. As we lately said, our memory has the power of lifting out of mere *subjectivism* into a certain knowledge of *objective* truth. Doubtless our faculties do not enable us to fully understand or know all the properties of bodies. It may well be that we might have additional senses which would reveal more of such properties—as we might have a sense which would enable us to perceive the magnetic qualities of bodies. But our perceptions, though inadequate to tell us all we might know, are not mendacious. We know that bodies exist objectively, and many of their objective conditions of existence, some of which give rise to two very notable abstract ideas, namely, those of *space* and *time*.

---

* See *ante*, p. 381. † See *ante*, p. 255.

All corporeal bodies are commonly spoken of as occupying some "portion of space," but we have no evidence of the existence of any such thing as "*space*," and the conception of ether* everywhere diffused, does away with all need for calling up such a phantom of the imagination. Things are "extended," but there is no such thing as "extension," and when we think of the latter, we always vaguely imagine some extended body, When we speak of "space," however, we do not mentally refer to any extended body; what we mean is the quality of extension, as completely abstracted from all bodies whatever and thought of purely by itself. "Extension" has, of course, no existence in itself *as* extension; though it is real and objective as a quality of real extended objects. "Space" is altogether ideal, an abstraction from abstractions, and when we speak of bodies "occupying space," we really refer to the exclusion of one extended body by another. "Space," then, though there is no such thing as space, is a true idea which signifies the extension of all extended things taken together and considered abstractedly. This truth does away with and explains the problem which has puzzled so many—is space endless or not? For it is not evident that there must be an endless succession of extended bodies in all directions, while space cannot go beyond extended bodies, since it is nothing but the abstract idea of their common extension and mutual exclusion. That we cannot *imagine* a boundary to space is simply due to our having had no such experience, for we can never imagine anything of which we have had no sort of experience; and we have

---

* See *ante*, p. 93.

had no experience of any extended body with nothing beyond it.

Similarly, "*time*" is an abstraction from abstractions. Things, in our experience, really endure more or less and succeed one another. But the ideas "endurance" and "succession," are nothing but abstractions in themselves, though they are real and objective as qualities of enduring and succeeding things. "Time" is altogether ideal; and when we speak of events as occurring in time, it is a mere mode of speaking which denotes the endurance and succession of succeeding things and the exclusion of one succeeding thing by another, together with the *duration* of the mutual exclusion of all succeeding things. In the absence then of any succession there could be no "time."

But the things which succeed and which are extended, we can, and do, know more or less perfectly, *as they are in themselves*, and apart from any distortion of truth due to the action of our own faculties. Thus, as beforesaid, two marbles *are* two, and would be two (we see and know) were every human mind annihilated, and would still be extended and rounded in shape were there no living creature on the surface of our planet.

We saw in the last chapter that the Greek philosophers, Arcesilaus and Carneades,* taught that we could know nothing but appearances, and that all our knowledge was merely relative. Unlike Pyrrho, they did not hesitate to affirm the truth of their systems of philosophy. The same system is professed by some amongst us to-day, and may be thus stated :

1. All our knowledge is merely relative.

---

* See *ante*, p. 330.

2. We can know nothing but appearances, *i.e.*, phenomena.

3. We cannot emerge from the sphere of objectivity and attain any knowledge of things objective.

Now of course anything which is "*known to us*," cannot at the same time be "*unknown to us*," and so far our knowledge may be said to affect the thing we know. But this is trivial. Our "knowing" or "not knowing" any object, is—apart from some act which may be the result of such knowledge—a mere accident of that object's existence, which is not otherwise affected thereby.

But indeed the system of the "relativity of knowledge" really refutes itself. For every system of knowledge must start with the assumption, implied or expressed, that something is true, or known so to be. By the teachers of the doctrine of the "relativity of knowledge," it is taught that the doctrine of the relativity of knowledge is thus true. But if we cannot know that anything corresponds with external reality, if nothing we can assert has more than a relative and phenomenal value, then this character must also appertain to the doctrine of the "relativity of knowledge" itself. Either, then, this system of philosophy is *itself* merely relative and phenomenal, and so cannot be known to be true, or else it is absolutely true, and can be known so to be. But it must be merely relative and phenomenal, if everything known by man is such. Its value then can be only relative and phenomenal, therefore it cannot be known to correspond with external reality, and cannot be asserted to be true. If anybody asserts that we can know the system of the relativity of knowledge to be true, he thereby asserts that it is false to say that all our knowledge is only relative. In asserting that we can know the system of the relativity of knowledge to be true, he who so asserts proclaims that some of our knowledge must be absolute; but this upsets the foundation of his whole system.

Any one who upholds such a system as this, may be compared to a man seated high up on the branch of a tree, which he is engaged in sawing across where it springs from the tree's trunk. The position and occupation of such a man can hardly be considered as evidence of wisdom on his part.

The simple but fundamental truths to which attention has been called in this chapter, are really present in the mind of every rational man, though they may be unobserved or very imperfectly apprehended by him. They have most important and far-reaching consequences, and merit the attention of every one who desires to be able to give a reasonable account of the convictions of his own mind. They are truths which are latent in, and so necessary for the validity of, every branch of science, that any *real* doubt about them would make not only experiment, but even observation, logically impossible.

The truths here put forward are, as before said, truths of science *par excellence*, as distinguished from its various branches with which we have been concerned in previous chapters. They are truths of *philosophy*, and as transcending every branch of "*physics*," constitute what we call "*metaphysics*."

But the student who, in perusing this chapter, meets with ideas and reasoning with which he was before unfamiliar, must not imagine that by mastering them he can acquire a knowledge of metaphysical truth or philosophy. Some fragments of that truth he can thus acquire, but he must bear in mind that in this chapter, as in the preceding chapters, we have but given him an introduction to the elements of the science therein treated of.

A philosophical foundation is absolutely necessary, however, to establish the validity of all sciences—that is, fundamental, or metaphysical, science must support and justify all its subordinate branches. By various combinations of the different lines of inquiry—different kinds of study—which we have considered in this book, each distinct science is constituted.

Thus (the co-operation of philosophy being in each case understood) by a combination of mathematics with the sciences which regard the stability and movements of bodies and the action of the various physical energies, we have the vast science of *Physics*. By the study of the laws of life, taken in conjunction with the lines of inquiry included under physics, we have *Biology*, with its subordinate group of sciences such as *physiology*, *anatomy*, &c. By the aid of psychology, taken with biology as applied to man, we have the special science of mankind or *Anthropology*, one department of which is logic. By the application of facts drawn from anthropology, especially as regards moral intuitions and moral relations, we have the science of right and wrong, that is *Ethics*. By a study of ethics and certain departments of anthropology, taken in conjunction with the law of causation and with history, we have *Theology*. Lastly, from anthropology, and the last-mentioned science, together with ethics, we may attain the science of, and rules for, the reasonable regulation of men in communities, so as to produce the greatest good, namely, the science and the art of *Politics*.

The object of this little book is to promote the cultivation of such organised knowledge. Its aim has been so to guide the student's first steps that, by a simultaneous introduction to the various sciences, he may come to apprehend the close connection and reciprocal relations which exist between them all. The author's hope is thus to stimulate desire for the pursuit of every one of those sciences the due cultivation of which is of such supreme importance to the welfare of mankind.

Printed by BALLANTYNE, HANSON AND CO.
London and Edinburgh

www.ingramcontent.com/pod-product-compliance
Lightning Source LLC
Chambersburg PA
CBHW051245300426
44114CB00011B/900